T0134418

Studies in Computational Intelligence

Volume 752

Series editor

Janusz Kacprzyk, Polish Academy of Sciences, Warsaw, Poland
e-mail: kacprzyk@ibspan.waw.pl

The series "Studies in Computational Intelligence" (SCI) publishes new developments and advances in the various areas of computational intelligence—quickly and with a high quality. The intent is to cover the theory, applications, and design methods of computational intelligence, as embedded in the fields of engineering, computer science, physics and life sciences, as well as the methodologies behind them. The series contains monographs, lecture notes and edited volumes in computational intelligence spanning the areas of neural networks, connectionist systems, genetic algorithms, evolutionary computation, artificial intelligence, cellular automata, self-organizing systems, soft computing, fuzzy systems, and hybrid intelligent systems. Of particular value to both the contributors and the readership are the short publication timeframe and the world-wide distribution, which enable both wide and rapid dissemination of research output.

More information about this series at http://www.springer.com/series/7092

Huimin Lu · Xing Xu
Editors

Artificial Intelligence and Robotics

 Springer

Editors
Huimin Lu
Department of Mechanical and Control
  Engineering
Kyushu Institute of Technology
Kitakyushu
Japan

Xing Xu
School of Computer Science
  and Engineering
University of Electronic Science
  and Technology of China
Chengdu
China

ISSN 1860-949X   ISSN 1860-9503   (electronic)
Studies in Computational Intelligence
ISBN 978-3-319-88856-9   ISBN 978-3-319-69877-9   (eBook)
https://doi.org/10.1007/978-3-319-69877-9

Printed on acid-free paper

This Springer imprint is published by Springer Nature
The registered company is Springer International Publishing AG
The registered company address is: Gewerbestrasse 11, 6330 Cham, Switzerland

Preface

In November 2017, the 2nd International Symposium on Artificial Intelligence and Robotics took place in Kitakyushu, Japan. This conference was organized by the International Society for Artificial Intelligence and Robotics (ISAIR, https://shinoceanland.com/conference/), Kitakyushu Convention and Visitors Association, Support Center for Advanced Telecommunications Technology Research Foundation, and Kyushu Institute of Technology, Japan. The annual organized series conferences focus on exchanges of the new ideas and new practices in industry applications. This book's objective is to provide a platform for researchers to share their thoughts and findings on various issues involved in artificial intelligence and robotics.

The integration of artificial intelligence and robotics technologies has become a topic of increasing interest for both researchers and developers from academic fields and industries worldwide. It is foreseeable that artificial intelligence will be the main approach to the next generation of robotics research. The explosive number of artificial intelligence algorithms and increasing computational power of computers has significantly extended the number of potential applications for computer vision. It has also brought new challenges to the artificial intelligence community. The aim of this book is to provide a platform to share up-to-date scientific achievements in this field. ISAIR2017 had received over 100 papers from over 11 countries in the world. After the careful review process, 35 papers were selected based on their originality, significance, technical soundness, and clarity of exposition. The papers of this book were chosen based on review scored submitted by members of the program committee and underwent further rigorous rounds of review.

It is our sincere hope that this volume provides stimulation and inspiration, and that it will be used as a foundation for works to come.

Kitakyushu, Japan
Chengdu, China
September 2017

Huimin Lu
Xing Xu

Acknowledgements

This book was supported by Leading Initiative for Excellent Young Research Program of Ministry of Education, Culture, Sports, Science and Technology of Japan (16809746), Grant-in-Aid for Scientific Research of JSPS (17K14694), National Natural Science Foundation of China (61602089), Research Fund of The Telecommunications Advancement Foundation, Fundamental Research Developing Association for Shipbuilding and Offshore, Strengthening Research Support Project of Kyushu Institute of Technology, Kitakyushu Convention and Visitors Association, and Support Center for Advanced Telecommunications Technology Research Foundation.

We would like to thank all authors for their contributions. The editors also wish to thank the referees who carefully reviewed the papers and gave useful suggestions and feedback to the authors. Finally, we would like to thank the Profs. Hyoungseop Kim, Seiichi Serikawa, Wen-yuan Chen, Junwu Zhu and all editors of Studies in Computational Intelligence for the cooperation in preparing the book.

About the Book

This edited book presents scientific results of the research fields of *Artificial Intelligence and Robotics*. The main focus of this book is on the new research ideas and results for the mathematical problems in robotic vision systems. The book provides an international forum for researchers to summarize the most recent developments and ideas in the field, with a special emphasis given to the technical and observational results obtained within recent years.

The chapters were chosen based on review scores submitted by editors and underwent further rigorous rounds of review. This publication captures 35 of the most promising papers, and we impatiently await the important contributions that we know these authors will bring to the fields of artificial intelligence and robotics.

Contents

Identification of the Conjugate Pair to Estimating Object Distance: An Application of the Ant Colony Algorithm

Shih-Yen Huang, Wen-Yuan Chen and You-Cheng Li

Abstract The 3D computer vision application become popular in recent years and estimating the object distance is basic technology. This study used laser array to beam the object then generate highlight characteristic point, and then applied Fuzzy C-mean (FCM) and Ant colony (ACO) to classify characteristic points on image. Finally, used conjugate pair and characteristic point on object and then based on Epipolar plane the object distance was estimated those maximum error rate is ±5.6%.

Keywords ANT colony · Conjugate pair · Laser array · Epipolar

1 Introduction

The 3D image computer vision technology application became popular, in the recent years. Particularly, the application in the Augmented Reality (AR) and 3D scan field. For example ATOS [1] 3D scan, as shown in Fig. 1 (cited from [1]), can identifying high resolution characteristic points on the object. The interval between the adjacent characteristic points have several types: 007, 0.12, 0.19 mm. The size of the object which be scanned from 185 × 140 to 500 × 380 mm. The minimum object distance is 440 mm.

S.-Y. Huang (✉) · Y.-C. Li
Department of Computer Science and Information Engineering,
National Chin-Yi University of Technology, Taichung, Taiwan, ROC
e-mail: syhuang@ncut.edu.tw

Y.-C. Li
e-mail: shen998826@yahoo.com.tw

W.-Y. Chen
Department of Electronic Engineering, National Chin-Yi University of Technology,
Taichung, Taiwan, ROC
e-mail: cwy@ncut.edu.tw

© Springer International Publishing AG 2018
H. Lu and X. Xu (eds.), *Artificial Intelligence and Robotics*,
Studies in Computational Intelligence 752,
https://doi.org/10.1007/978-3-319-69877-9_1

1

Fig. 1 ATOS core 3D scan
(cited from [1])

Fig. 2 HANDYSCAN 300
(cited from [2])

The other high performance product is HANDYSCAN 300 [2], as shown in Fig. 2 (cited from [2]), which is manufactured by Creaform3d company. HANDYSCAN 300 can process 205,000 characteristic points and interval only 0.04 mm between adjacent characteristic points. Each scan area is 225 × 250 mm. The nearest object distance is 300 mm. In order to compatible with various 3D systems software, this scanner output data have plentiful file type such as:.dae,.fbx,.ma,.obj,.ply,.stl,.txt,.wrl,.x3d,.x3dz,.zpr.

In recent year, many researcher is interested to applied machine learning technique for computer vision. For example, Lu et al. [3] used artificial life with an image function. Xu et al. [4] proposed a method to decrease the quantization loss. Lu et al. [5] proposed a method to alleviate the intensity in homogeneity and color distortion. To accurately learn marine organisms, Lu et al. [6] proposed FDCNet. However, estimating the object distance is the basic technology of 3D scan or 3D computer vision. This paper proposed a simple method that used laser array to highlight the characteristic points on the object, used two parallel camera to capturing objects then shown on left and right image, applied Epipolar plane [7] to estimating the distance between lens center and the characteristic point on the object.

2 Scheme of Estimating Object Distance

(1) Applied Epipolar plane to estimate object distance

Object is captured by two parallel cameras, an Epipolar plane is constructed by each point of the object and the focus centers of the two camera [7] as shown in Fig. 3. Where, Point O, A is the principle focus of left camera and right camera respectively. Point P is a point on the object. The line segment $\overline{OD}$ and $\overline{AF}$ is focal length of the left and right camera respectively. Point D and F is the center pixel on the left and right camera respectively. Point P is shown as pixel E and G on the left and right image respectively. In other word, pixel E and G are the conjugate pair [7] on this Epipolar plane. Obviously the length of $\overline{DE}, \overline{FG}, \overline{BP}, \overline{CP}$ can be measured. The two triangle ΔPBO and ΔEDO are similar. ΔPCA and ΔGFA are similar also. The object distance $\overline{DB}$ can be solved by the following Eqs. (1) and (2).

$$\frac{\overline{BP}}{\overline{DE}} = \frac{\overline{OB}}{\overline{OD}} \tag{1}$$

$$\frac{\overline{CP}}{\overline{FG}} = \frac{\overline{AC}}{\overline{AF}} \tag{2}$$

(2) Highlight the point on the object

Since the conjugate pair, e.g. point E and G, are identified by image processing technology. If the feature of point P on the object is hard to identified, the conjugate pair is fail [8]. Accordingly, we used laser array to beaming the object thus these highlight points on the object were reflected as point array. The green points in Fig. 4 denoted the points on the image those were the points on the object beamed by laser and the black points denoted the pixels those were the image captured from part of the object where no laser beaming on. Obviously, applying traditional image processing technology these green points are identified easily. The experimental results shown the conjugate pair were constructed easily.

Fig. 3 Epipolar plane is constructed by two parallel camera

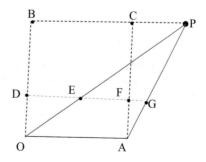

S.-Y. Huang et al.

Fig. 4 The green points are
the points those were beamed
by laser array

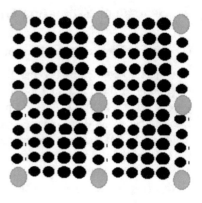

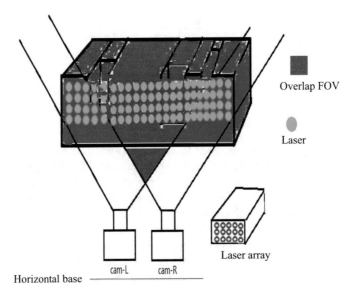

Fig. 5 System architecture included two cameras and laser array

(3) System architecture

In this study, we used two parallel cameras and laser array to beaming the objects. In Fig. 5, the red area denoted the overlap FOV (Field of View) of the left and right camera. The green points denoted the points beamed by the laser array. The cam-L, cam-R denoted the left and right camera respectively.

3 Identified Characteristic Point

The laser point shown on the left and right image is the characteristic point, named left K-point and right K-point those composed a conjugate pair as described previously. However, each K-point is image shown the red laser points array on flat object, which contain many color and pixels as shown in Fig. 6. Based on image classification to identify the K-points and calculate each centroid, and then each conjugate pairs are constructed each Epipolar plane. Finally objects distance between the lase point and the lens can be estimated by Eqs. (1) and (2). The block diagram of this processing were shown in Fig. 7. Furthermore, using optimal algorithm to classify image is popular method. In this paper, we applied two optimal algorithm: Fuzzy C-mean and Ant colony method to classify image.

(a)   **(b)**

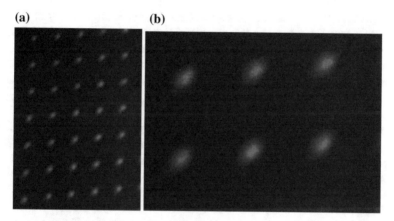

Fig. 6 The red objects in the image are the laser points on the object those captured by camera. **a** The image of the red laser point array. **b** Enlarged some laser points

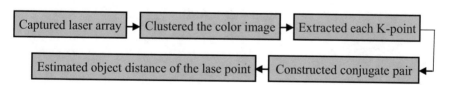

Fig. 7 The block diagram to estimate the object distance

Fig. 8 Illustrated three
objects with different high

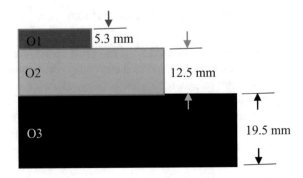

4 Experimental Results

To estimate the object distance, we selected three objects (named as object O1, O2 and O3) which with various high as illustrated in Fig. 8. High of object O1 (red), O2 (green) and O3 (black) is 5.3 mm, 12.5 mm and 19.5 mm respectively. The captured left, right image are the three objects shown in Fig. 9a and d, respectively.

The captured left and right image is these objects beamed by laser array as shown on Fig. 9b and e respectively. Applied Ant colony algorithm to classify these image, and then the K-points were extracted from the classified image was shown in Fig. 9c and f respectively.

Each laser point on the object and the corresponding K-points on the left and right image those are on the Epipolar plane. According to Eqs. (1) and (2) the object distance of each laser point can be estimated. Finally, the high of each objects can be calculated from these object distances. Table 1 illustrated the experimental results. In Table 1, High was measured by vernier caliper, High_FCM and High_ACO was estimated based on FCM and Ant colony respectively. Experimental results shown that applied Ant colony to classify laser point can estimate object distance was more accurate than FCM did. This study defined error rate, *err*, as Eq. (3) described. The experimental results shown that the maximum error rate is ±5.6%.

$$err = \frac{High - High\_ACO}{High} \times 100\% \tag{3}$$

5 Conclusions

This study used laser array to beam object to highlight the characteristic points. Accordingly, could simply identified the characteristic points on the Epipolar plane which could estimate the object distance of these characteristic points on the object. In addition, this study applied FCM and Ant colony algorithm to classify these

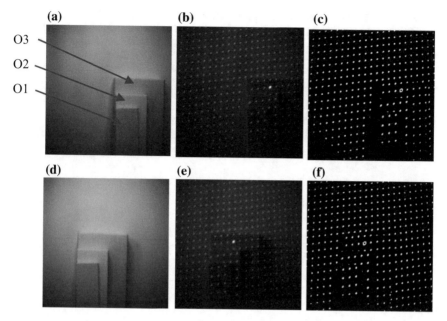

Fig. 9 The images had three objects those with different high. **a** Left image which was captured by left camera. **b** Left Image shown objects were beamed by laser array. **c** Extracted the K-points from left image. **d** Right image which was captured by left camera. **e** Right Image shown objects were beamed by laser array. **f** Extracted the K-points from right image

Table 1 Estimated high of the object, where the unit is millimeter

Object	High	High_FCM	High_ACO	Err %
O1	5.3	4	5.1	3.8
O2	12.5	13.5	13	−4
O3	19.5	20.5	20.6	−5.6

object images those were captured by the left and right camera. The experiment result shown that the high of the object can be estimated within ±5.6% error rate.

References

1. http://www.rat.com.tw/. Accessed 12 July 2017
2. https://www.creaform3d.com/en/metrology-solutions/. Accessed 12 July 2017
3. Lu, H., Li, Y., Chen, M., Kim, H., Serikawa, S.: Brain intelligence: go beyond artificial intelligence. Int. J. Comput. Sci.—Comput. Vis. Pattern Recogn. (2017) arXiv:1706.01040
4. Xu, X., He, L., Shimada, A., Taniguchi, R.-I., Lu, H.: Learning unified binary codes for cross-modal retrieval via latent semantic hashing. Neurocomputing **213**, 191–203 (2016)

5. Lu, H., Li, B., Zhu, J., Li, Y., Li, Y., Xu, X., He, L., Li, X., Li, J., Serikawa, S.: Wound intensity correction and segmentation with convolutional neural networks. J. Concurr. Comput. Pract. Exp., **29**(6) (2017). http://onlinelibrary.wiley.com/doi/10.1002/cpe.v29.6/issuetoc
6. Lu, H., Li, Y., Uemura, T., Ge, Z., Xu, X., He, L., Serikawa, S.: FDCNet: filtering deep convolutional network for marine organism classification. J. Multimed. Tools Appl., 1–14 (2017)
7. Chung, K.L.: Image Processing And Computer Vision, Tunghua, Inc., ISBN 978-957-483-827-1 (2015)
8. Huang, S.Y., Chen, Y.C., Wang, J.K.: Measurement of tire tread depth with image triangulation. In: Proceedings of 2016 International Symposium on Computer, Consumer and Control (IS3C 2016), July 04–06, Xi'an, China, pp. 303–306 (2016)

Design of Palm Acupuncture Points Indicator

Wen-Yuan Chen, Shih-Yen Huang and Jian-Shie Lin

Abstract The acupuncture points are given acupuncture or acupressure so to stimulate the meridians on each corresponding internal organs with a treatment of physical illness. The goal of this study is to use image technique to automatically find acupuncture positions of a palm to help non-related professionals can clearly identify the location of acupuncture points on him palm. In this paper, we use the skin color detection, color transform, edge detection, histogram and fast packet method to extract the palm and find out the acupunctures. First, we use fast packet method to get the acupunctures of the finger. And then a histogram technique was used to obtain the acupuncture points of the valley of the fingers. Finally, the valley points and fingertips of the finger are used as a reference combined with the standard deviation of data images to calculate the position of the palm acupuncture points. From the simulation result, it is demonstrated that our design is an effective method for indicating the acupuncture points of a palm.

Keywords Acupuncture · Image recognition · Finger valley
Palm · Histogram

1 Introduction

For finding the acupuncture points of a palm, a palm extraction from image or hand gesture method and acupuncture knowledge are necessary. Mazumdar et al. [1] published a hand gesture detection method for human and machine interaction. Its

W.-Y. Chen · S.-Y. Huang (✉) · J.-S. Lin
Department of Electronic Engineering, National Chin-Yi University of Technology Taiping, Taichung 41170, Taiwan (R.O.C.)
e-mail: syhuang@ncut.edu.tw

W.-Y. Chen
e-mail: cwy@ncut.edu.tw

J.-S. Lin
e-mail: h647376@gmail.com

© Springer International Publishing AG 2018
H. Lu and X. Xu (eds.), *Artificial Intelligence and Robotics*,
Studies in Computational Intelligence 752,
https://doi.org/10.1007/978-3-319-69877-9_2

goal is focus on the video hand gesture recognition. In this research, they estimate the performers of the hand tracking and announce a hand extraction scheme by wear the glove. This method can handle the finger at part by part manner. Throughout the tracking, the hand still motion freely. Besides, the system can move the noise effectively and working well in complex background.

Duan et al. [2] research the hand gesture and human-machine interaction deeply. It provide a core for accuracy and fast to identify the hand gesture in motion object. Meanwhile, they also propose a new building mode of dynamic and real time hand gesture extraction technologies. First, they check the rate variation of motion to confirm the start and stop of the dynamic hand gesture. And then a mean-drift algorithm associated with the target detection of the hand gesture motion and color information to achieve the real time tracking. Experimental results reveal the method reach the real time and stability advantages.

Causon et al. [3] use multiple camera as the Fig. 1 shown to extract hand image and capture many pictures on different angular. The system is used to extract data to construct the hand model by 3D image structure. From simulation result, it demonstrates the 3D structure is a well design for hand gesture detection. About the other hand extraction and gesture research can be found in [4–6].

Yang et al. [7] proposed a method use data mining technique to analyze the effects of treatment and the influence of behavioral variables by using fifty patients with juvenile myopia. On acupuncture treatment, myopia patients were divided into two classes for clustering analysis. From the experiments result, it is demonstrate that a good treatment of acupuncture could slow the progression of juvenile myopia.

Birch and Hammerschlag [8] release a future issues of clinical acupuncture and oriental medicine include four topics: designing clinical research to evaluate traditional East Asian system of medicine; identifying and controlling for the non-specific effects of acupuncture; assessing strength and weakness of systematic reviews of acupuncture; identifying research questions relevant to the acupuncture

Fig. 1 The schematic of the multiple camera structure for hand gesture extraction

community. Those topics are plain to describe the clinical acupuncture is good for the human health. About the other acupuncture research can be found in [9, 10].

Using artificial lift with an image function Lu et al. [11] proposed Brain Intelligence (BI) for the events without having experienced. To gain the binary code learning, Xu et al. [12] used class label and proposed a method to decrease the quantization loss. To increase the computing efficiency, Lu et al. [13] used an additive operator-splitting algorithm and proposed model-based method to alleviate the intensity inhomogeneity. Lu et al. [14] proposed Filtering Deep Convolutional Network (FDCNet) and shown this method was better than state-of-the-art classification method.

2 System Algorithm

In this study, we enter a test image with hand message and then verify if the image has a palm. If the answer is no then the system loop to the enter image stage otherwise go to the skin detection stage. On skin detection, a color transformation is

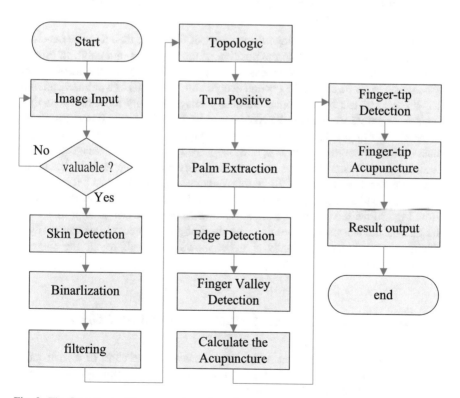

Fig. 2 The flow chart of the acupuncture of a palm

hired to map a color image from RGB space to YCbCr space. Then a set of equations was used to identify the pixels are belong to the skin color or not. Due to binary image is suitable for the feature detection; we convert the color image into the binary image. Next, a filter use to filter the noise out. A topologic method is used to patch up the object. Certainly, turn positive stage is necessary for providing to obtain a well detection. Successively, a palm extraction method was used to distinguish the palm portion. Then, edge detection extracts the edge of the image. It is provided the system to find the valley of the finger. It is well known, the valley of the finger is a location of the acupuncture. In this research, the acupunctures of the palm were obtained by calculate the position according to the database and acupunctures of the valley. As for the fingertip, we used a fast packet method associate with the edge of a palm image can easily to get. Finally, we add the all acupunctures, the acupunctures of a palm was completed. The details of the above mention can be found in Fig. 2.

3   Experiment

3.1   Acupuncture Location Calculation of a Fingertip

The acupuncture location of fingertip is a part of the palm; therefore we develop a fast packet method to calculate the position of the fingertips in a palm. In order to describe the details of finding process, we use Fig. 3 to explain the theoretic and operation. Figure 3a shows the general acupuncture positions of a palm. Meanwhile it is an original test image. Since the binary image is easily to extract something features, therefore we transfer the color image into binary image. Figure 3b display the result after binarlized corresponding to the Fig. 3a. It is well known the fingertip must be on the contour of an image. Since a method to find the image contour is necessary, we transfer the color image to the binary image. Figure 3c is the contour image that is extract from Fig. 3b. In this research, we develop a fast packet method to get the tips of a finger. Figure 3d shows the schematic of the fast packet method. From Fig. 3d we see the tips are all on the intersection between contour image and the line on the fast packet. Certainly, by the fast packet method, we can easily to obtain all the fingertips. Finally, Fig. 3e shows the result of all the fingertip acupuncture points from a palm.

3.2   Acupuncture Location Calculation of a Man Palm

Figure 4 shows the case of an acupuncture location calculation of a man palm. Figure 4a shows the result of a fast packet method when it is applied to fingers of a palm. This method can fast to find out the tips of fingers from a palm. In other

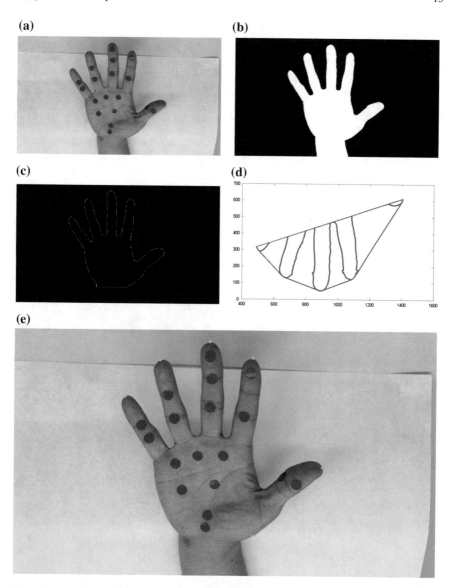

Fig. 3 The tip of finger grab case: **a** the original image, **b** the result after binarization, **c** the contour of a palm, **d** the schematic of a fast packet method **e** the result of a tip of finger grabbed

words, we can easily use the fast packet method to extract the acupuncture location of the tip of fingers. Figure 4b shows result of the acupuncture location of a fingertip. Where, the green points are the acupuncture points of fingertip of a palm. Figure 4c is the histogram of a palm corresponding to the Fig. 4b. We use histogram technique to find out all the valleys that it is map to a finger of the palm. It is

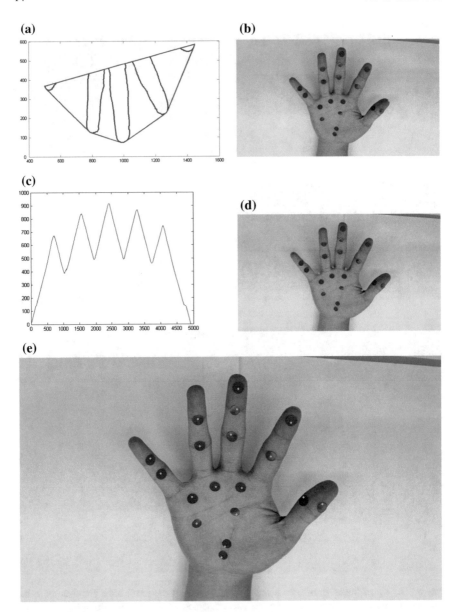

Fig. 4 The location calculation of a man palm case: **a** the result of a fast packet method, **b** the acupuncture location of a fingertip, **c** the histogram of a palm, **d** the acupuncture location of valley of finger, **e** the calculation result of acupuncture of a palm

reveals that the each local valley is corresponding to the each valley of the fingers. This method can exactly to find the valleys of a palm. Figure 4d expresses the acupuncture location of valley of finger; it is corresponding to the Fig. 4c. From the picture, we can easily to see the acupuncture points; the green points display. Finally, Fig. 4e shows the calculation result of acupuncture of a palm. The same as the Fig. 4b, d shown, the green points is specially use to display the acupuncture location. It is design for peoples to easily to watch. The details can be found in Fig. 4, please to watch it carefully.

4   Conclusion

This study uses the features of the tip and valley of the finger to find out the all acupuncture points of itself of the finger. Meanwhile, we also calculate all the acupuncture points that are located in the palm. In the experiment, the skin color used to extract the palm image. The filter used to filter out the noise to obtain a suitable successive detection image. The topologic technique used to fill out the image for further edge detection. Certainly, the principle axis method used to turn the image into positive position. Finally, the histogram and fast packet method are adopted to extract the five tips and four valleys acupuncture of the fingers. By the way, according to find out acupunctures and database, we calculate all the acupunctures of a palm. From the simulation result, it reveals our system is an effective and corrective method for pointing all the acupuncture points of a palm.

References

1. Mazumdar, D., Talukdar, A.K., Sarma, K.K.: Gloved and free hand tracking based hand gesture recognition. In: ICETACS 2013, pp. 197–202 (2013)
2. Duan, H., Zhang, Q., Ma, W.: An approach to dynamic hand gesture modeling and real-time extraction. In: IEEE conference, pp. 139–142 (2011)
3. Causo, A., Matsuo, M., Ueda, E., Takemura, K., Matsumoto, Y., Takamatsu, J., Ogasawara, T.: Hand pose estimation using voxel-based individualized hand model. Adv. Intell. Mech. 451 456 (2009)
4. Zhang, H., Zhu, Q., Guan, X.-F.: Probe into image segmentation based on Sobel operator and maximum entropy algorithm. In: IEEE 2012 International Conference on Computer Science and Service System, pp. 238–241 (2012)
5. Zhao, Y., Zhang, D., Wang, Y.: Automatic location of facial acupuncture-point based on content of infrared thermal image. In: 2010 5th International Conference on Computer Science and Education (ICCSE), pp. 65–68 (2010)
6. Ren, X.: RGB change their perception: RGB-D for 3-D modeling and recognition. IEEE Robot. Autom. Mag. **20**(4), 49–59 (2013)
7. Yang, X., Xu, L., Zhong, F., Zhu, Y.: Data mining-based detection of acupuncture treatment on juvenile myopia. J. Tradit. Chin. Med. **32**(3), 372–376 (2012)
8. Birch, S., Hammerschlag, R.: International workshop on acupuncture research methodology papers on Topic 1: Matching research design to research question in clinical trials of acupuncture. Clin. Acupunct. Orient. Med. **3**(4), 191 (2002)
9. Cross, J.R.: Modern Chinese acupuncture—A review of acupuncture techniques as practised in China today. Physiotherapy **81**(5), 302 (1995)

10. Baxter, G.D.: Acupuncture in clinical practice—A guide for health professionals (Therapy in Practice 43). Physiotherapy **81**(11), 701 (1995)
11. Lu, H., Li, Y., Chen, M., Kim, H., Serikawa, S.: Brain intelligence: go beyond artificial intelligence. Int. J. Comput. Sci. Comput. Vision Pattern Recogn. (2017). https://arxiv.org/abs/1706.01040
12. Xu, X., He, L., Shimada, A., Taniguchi, R.-I., Lu, H.: Learning unified binary codes for cross-modal retrieval via latent semantic hashing. Neurocomputing **213**, 191–203 (2016)
13. Lu, H., Li, B., Zhu, J., Li, Y., Li, Y., Xu, X., He, L., Li, X., Li, J., Serikawa, S.: Wound intensity correction and segmentation with convolutional neural networks. J. Concurrency Comput. Pract. Exp. **29**(6) (2017). http://onlinelibrary.wiley.com/doi/10.1002/cpe.v29.6/issuetoc
14. Lu, H., Li, Y., Uemura, T., Ge, Z., Xu, X., He, L., Serikawa, S.: FDCNet: filtering deep convolutional network for marine organism classification. J. Multimedia Tools Appl. 1–14 (2017)

Low-Rank Representation and Locality-Constrained Regression for Robust Low-Resolution Face Recognition

Guangwei Gao, Pu Huang, Quan Zhou, Zangyi Hu and Dong Yue

Abstract In this paper, we propose a low-rank representation and locality-constrained regression (LLRLCR) based approach to learn the occlusion-robust discriminative representations features for low-resolution face recognition tasks. For gallery set, LLRLCR uses double low-rank representation to reveal the underlying data structures; for probe set, LLRLCR uses locality-constrained matrix regression to learn discriminative representation features robustly. The proposed method allows us to fully exploit the structure information in gallery and probe data simultaneously. Finally, after getting the resolution-robust features, a simple yet powerful sparse representation based classifier engine is used to predict the face

G. Gao (✉) · P. Huang · D. Yue
Institute of Advanced Technology, Nanjing University
of Posts & Telecommunications, Nanjing, China
e-mail: csgwgao@njupt.edu.cn

P. Huang
e-mail: huangpu@njupt.edu.cn

D. Yue
e-mail: yued@njupt.edu.cn

G. Gao · Q. Zhou
Key Laboratory of Intelligent Perception and Systems for High-Dimensional
Information of Ministry of Education, Nanjing University of Science & Technology,
Nanjing, China
e-mail: quan.zhou@njupt.edu.cn

Q. Zhou
National Engineering Research Center of Communications and Networking,
Nanjing University of Posts & Telecommunications, Nanjing, China

Z. Hu
School of Automation, Nanjing University of Posts & Telecommunications,
Nanjing, China
e-mail: Huzangyi@163.com

© Springer International Publishing AG 2018
H. Lu and X. Xu (eds.), *Artificial Intelligence and Robotics*,
Studies in Computational Intelligence 752,
https://doi.org/10.1007/978-3-319-69877-9_3

labels. Experiments conducted on the AR database with occlusions have shown that the proposed method can obtain promising recognition performance than many state-of-the-art LR face recognition approaches.

Keywords Low-resolution · Face recognition · Low-rank · Nuclear norm Locality

1 Introduction

Face recognition (FR) has been an active research field for more than two decades and many promising face recognition methods have been developed [1–10]. However, due to the long distance to the interest object, the captured face images are usually low-resolution (LR) and lacking detailed facial features for face recognition. Generally, there are three standard approaches to match a LR probe image to a high-resolution (HR) gallery of enrolled faces: (i) down-sample the HR gallery image to match the LR probe one; (ii) up-sample the LR probe image to match the HR gallery; (iii) reduce the gap between different resolution and then apply the traditional face recognition methods on the common space. Although down-sampling alleviates the mismatch problem intuitively and easily, much of the available information may be lost during down-scaling.

Super-resolution (SR) is widely used to up-sample the LR probe images to fit the resolution gap with HR gallery ones. Recently, many techniques of decomposing a complete image into smaller patches have been introduced. Chang et al. [11] presented a neighbor embedding (NE) based super-resolution method. Most recently, many researchers state that the position information is very important for face analysis and synthesis. Ma et al. [12] proposed a position-patch based face hallucination method using all patches from the same position in a dictionary. To overcome the unstable solution of least square problem in [12], Jung et al. [13] employed sparsity prior to improve the reconstruction result with several principal training patches. Recently, Jiang et al. [14] incorporated a locality constraint into the least square inverse problem to maintain locality and sparsity simultaneously.

These vision-based super-resolution methods aim at achieving good image quality, and they do not consider discriminative features which limit the face recognition performance. Resolution robust feature based algorithms do aim at improving the recognition rate instead of the visual quality. Hennings-Yeomans et al. [15] proposed a simultaneous super-resolution and feature extraction method to express the constraints between LR and HR images in a regularization formulation. Biswas et al. [16] proposed to use multidimensional scaling (MDS) to embed LR images in a distance-preserving feature space together with the HR ones. Jiang et al. [17] proposed a coupled discriminant multi-manifold analysis (CDMMA) based LR face recognition method. CDMMA learnt mapping simultaneously from the neighboring information as well as the local geometric structure implied by the samples. All these algorithms perform resolution-robust feature extraction for classification. However, due to the degrading process from HR to LR space, LR images have limited discriminability both in reconstruction and recognition. Lu et al. [18] proposed a semi-coupled locality-constrained representation (SLR) approach to learn the discriminative representations and the mapping relationship between LR and HR features simultaneously.

However, they ignore the occlusions in the LR probe when learning the representation features. To handle LR face recognition with occlusions, we propose a novel method called low-rank representation and locality-constrained regression (LRRLCR) in this paper. Here, LLRLCR aims at using double low-rank representation to reveal the underlying data global structures of gallery set. Also, LLRLCR uses locality-constrained matrix regression to learn robust and discriminative representation features of the probe data. After getting the resolution-robust representation features both in LR and HR image space, a simple yet powerful sparse representation based face recognition engine is used to predict the label information. Experimental results on the AR database with occlusions show the superiority of our proposed method over some existing LR face recognition algorithms.

2 The Proposed LLRLCR

2.1 Double Low-Rank Representation

For the gallery data $X = [x_1, x_2, ..., x_n] \in \Re^{d \times n}$, we have:

$$\min_{Z, E_i} \|Z\|_* + \lambda \sum_{i=1}^{n} \|E_i\|_*, \quad s.t. \quad X = DZ + E \tag{1}$$

where $D \in \Re^{d \times m}$ is the given dictionary, $\| \cdot \|_*$ is the nuclear norm (i.e. the sum of the singular values) of a matrix. Each error $E_i \in \Re^{p \times q}$ is a matrix $(d = pq)$.

Inexact Augmented Lagrange multiplier (IALM) method [19] is used to solve model (1). We first convert model (1) to the following equivalent problem:

$$\min_{Z, E_i, J} \|J\|_* + \lambda \sum_{i=1}^{n} \|E_i\|_*, \quad s.t. \quad X = DZ + E, \ Z = J \tag{2}$$

The augmented Lagrange function L is given by:

$$L_\mu(Z, E_i, J, Y_1, Y_2) = \|J\|_* + \lambda \sum_{i=1}^{n} \|E_i\|_* + <Y_1, X - DZ - E>$$
$$+ <Y_2, Z - J> + \frac{\mu}{2}\left(\|X - DZ - E\|_F^2 + \|Z - J\|_F^2\right) \tag{3}$$

where $<A, B> = trace\,(A^T B)$, Y_1 and Y_2 are the Lagrange multiplier, $\mu > 0$ is a penalty parameter. The whole procedure is outlined in Algorithm 1. Step 1 and 3 can be solved by the singular value thresholding operator [20].

2.2 Locality-Constrained Matrix Regression

To better reveal the occlusion in the testing data, we use nuclear norm to characterize the reconstruction error. Our model can be formulated as follows:

$$\min_{x} \|y - D(x)\|_* + \lambda\|d \otimes x\|_2^2 \tag{4}$$

where $d = (d_1, \ldots, d_N)^T$ is distance vector, $D(x) = x_1 D_1 + x_2 D_2 + \cdots + x_m D_m$, each $D_j \in \Re^{p \times q}$ $(j = 1, \ldots, m)$ is a matrix, and $d_i = \|y{-}D_i\|_F^2$ is the locality metric. We rewrite the optimization problem of (4) as:

$$\min_{x, E} \|E\|_* + \lambda\|d \otimes x\|_2^2, \quad s.t. \quad y - D(x) = E, \tag{5}$$

The above problem can be solved via the alternating direction method of multipliers (ADMM) [9, 10] with the following augmented Lagrangian function:

$$L_\mu(x, E) = \|E\|_* + \lambda\|d \otimes x\|_2^2$$
$$+ Tr\left(Y_3^T(y - D(x) - E)\right) + \frac{\mu}{2}\|y - D(x) - E\|_F^2 \tag{6}$$

where Y_3 is the Lagrange multiplier, μ is a penalty parameter. The whole procedure is outlined in Algorithm 2. Step 2 can be solved by the singular value thresholding operator [20].

Algorithm 1. Solving problem (1) via IALM

Input: Training datasets D, gallery datasets X, regularization parameter λ.

Initialize: $J = Z = Y_2 = 0 \in \Re^{m \times n}$, $E = Y_1 = 0 \in \Re^{d \times n}$, $\rho = 1.1$, $\mu = 10^{-6}$, $max_\mu = 10^{10}$, $\varepsilon = 10^{-7}$.

---

while not converged **do**

  1: Fix the others and update J by

$$J = \arg\min\left(\frac{1}{\mu}\|J\|_* + \frac{1}{2}\|J - (Z + Y_2/\mu)\|_F^2\right)$$

  2: Fix the others and update Z by

$$Z = \left(I + D^T D\right)^{-1}\left(D^T X - D^T E + J + \left(D^T Y_1 - Y_2\right)/\mu\right)$$

  3: Fix the others and update E_i by

$$E_i = \arg\min\frac{\lambda}{\mu}\|E_i\|_* + \frac{1}{2}\left\|E_i - \left(X_i - D(Z_i) + Y_1^i/\mu\right)\right\|_F^2$$

  where $D(Z_i) = z_{1i} D_1 + z_{2i} D_2 + \ldots + z_{mi} D_m$, each D_j ($j=1,\ldots,m$) is a matrix.

  4: Update the multipliers:

$$Y_1 = Y_1 + \mu\left(X - DZ - E\right)$$
$$Y_2 = Y_2 + \mu\left(Z - J\right)$$

  5: Update the parameter μ by $\mu = \min(\rho\mu, max_\mu)$

  6: Check the convergence conditions:

$$\max\left(\|X - DZ - E\|_\infty, \|Z - H\|_\infty\right) < \varepsilon$$

end while

Output: Representation features Z.

Algorithm 2. Solving problem (4) via ADMM

Input: Training datasets D, probe input y, regularization parameter λ.

Initialize: $E = Y_3 = 0 \in \Re^{p \times q}$.

---

while not converged **do**

  1: Fix the others and update x by $x - (G + \tau B^2)\backslash ones(m,1)$, where $\tau = 2\lambda/\mu$, B is a $m \times m$ diagonal matrix with entries $B_{jj} = d_j$, and G is the covariance matrix $G = C^T C$ with $C = (y - E + Y_3/\mu)ones(m,1)^T - H$.

  2: Fix the others and update E by

$$E = \arg\min\left(\frac{1}{\mu}\|E\|_* + \frac{1}{2}\|E - (y - D(x) + Y_3/\mu)\|_F^2\right)$$

  3: Update the multipliers:

$$Y_3 = Y_3 + \mu\left(y - D(x) - E\right)$$

  4: Check the convergence conditions: $\|y - D(x) - E\|_\infty < \varepsilon$

end while

Notice: $H = [Vec(D_1), Vec(D_2),\ldots, Vec(D_m)]$.

Output: Coefficient vector x.

2.3   Classification

The low-rank representation matrix Z of HR gallery dataset X_h over the HR training dataset D_h can be computed by Eq. (1). For the LR probe y, its locality-constrained coefficient x_y over the LR training dataset D_l can be computed by Eq. (4). Then the sparse coding coefficient can be indicated as follows:

$$\min_{w} \ \left\| x_y - Zw \right\|_2^2 + \eta \|w\|_1 \tag{7}$$

where η is the balance parameter. After solving problem (7), the optimal w^* can be obtained. Then, the class-wise reconstruction error can be calculated through

$$e_i(x_y) = \left\| x_y - Z_i \delta_i(w^*) \right\|_2^2 \tag{8}$$

where δ_i is a function that selects the coefficients associated with the i-th class, and the label of the input y is assigned to the class, whose reconstruction error is the smallest.

3   Experimental Results and Discussions

3.1   Dataset Description

We use The AR database [21] to evaluate the performance of the proposed algorithm. The database contains 2,600 face images for 100 people, including frontal views of faces with different facial expressions, lighting conditions and occlusions. For each subject, twenty-six face images are taken in two separate sessions. For each subject, we choose seven clean images from the first session to form training sets D_h and D_l, seven clean images from the second session to form HR gallery set X_h, and six images with sunglasses and scarfs from the first session to form LR probe set Y_l. Therefore, all the probe images are not in the gallery set. We set HR face image as 44 × 32 pixels and down-sampling the HR by scaling factor of 4. The LR image is at 11 × 8 pixels. Some samples are shown in Fig. 1.

3.2   Comparison with the State-of-the-Art Methods

In this section, we will fully compare the proposed method with some state-of-the-art face recognition algorithms: one is hallucination-based face recognition methods using hallucinated HR images as testing inputs, such as Bicubic interpolation (BIC), Ma et al.'s [12] position patch based super-resolution (LSR),

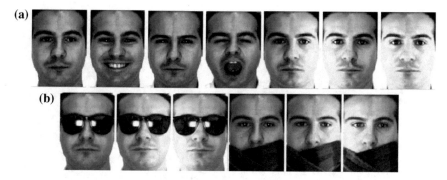

Fig. 1 Samples from AR face database: **a** Gallery set; **b** Probe set

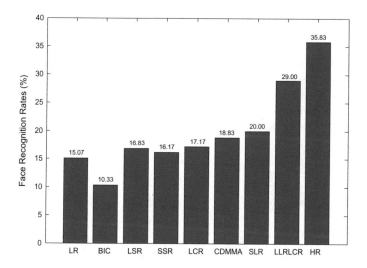

Fig. 2 Recognition rates of different methods on AR face database

Jung et al.'s [13] sparse representation based super-resolution (SSR) and Jiang et al.' [14] locality-constrained representation based super-resolution (LCR). They all use SRC [22–24] for recognition tasks. The others are typical resolution-robust LR face recognition methods just using LR images as testing inputs, which include CDMMA [17] and SLR [18]. All these experiments are based on the same gallery dataset and probe dataset for face recognition algorithms and the same training set for hallucination algorithms.

The recognition rates are reported in Fig. 2. We can observe that: (1) The hallucination based methods obtain unsatisfactory recognition rates. The reason may be that, due to the occlusions in LR probe image, these SR methods cannot well learn the representation coefficients, result in the loss of discriminative details in the hallucinated images. Some SR results are listed in Fig. 3 for better illustration;

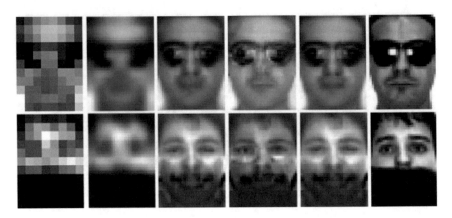

Fig. 3 Hallucinated results based on different methods. From left to right are the Input LR image, the results of BIC, LSR [12], SSR [13], LCR [14] and the original HR image

(2) The proposed method significantly improves the performance when compared to the hallucination based approaches and outperforms CDMMA and SLR. These results confirm that by introducing the low-rank representation and locality-constrained matrix regression, our proposed LLRLCR can learn more robust and discriminative representation features, which boost the face recognition performance.

4 Conclusions

In this paper, we have proposed a low-rank representation and locality-constrained regression (LRRLCR) based model to improve the discriminability of representation features of LR images with occlusions. LRRLCR aims at revealing the underlying structure information of the gallery data and learning robust and discriminative representation features for recognition tasks simultaneously. Experiments on LR face recognition methods have been conducted on AR face dataset with occlusions, confirming the superiority of our proposed methods over several state-of-the-art approaches.

Acknowledgements This work was partially supported by the National Natural Science Foundation of China under Grant nos. 61502245, 61503195, 61401228, the Natural Science Foundation of Jiangsu Province under Grant no. BK20150849, the China Postdoctoral Science Foundation under Grant no. 2016M600433, Open Fund Project of Key Laboratory of Intelligent Perception and Systems for High-Dimensional Information of Ministry of Education (Nanjing University of Science and Technology) (No. JYB201709 and JYB201710).

References

1. Jing, X.-Y., Wu, F., Zhu, X., Dong, X., Ma, F., Li, Z.: Multi-spectral low-rank structured dictionary learning for face recognition. Pattern Recogn. **59**, 14–25 (2016)
2. Lai, Z., Wong, W.K., Xu, Y., Yang, J., Zhang, D.: Approximate orthogonal sparse embedding for dimensionality reduction. IEEE Trans. Neural Netw. Learn. Syst. **27**, 723–735 (2016)
3. Wang, D., Lu, H., Yang, M.H.: Kernel collaborative face recognition. Pattern Recogn. **48**, 3025–3037 (2015)
4. Yang, W., Wang, Z., Sun, C.: A collaborative representation based projections method for feature extraction. Pattern Recogn. **48**, 20–27 (2015)
5. Gao, G., Yang, J., Wu, S., Jing, X., Yue, D.: Bayesian sample steered discriminative regression for biometric image classification. Appl. Soft Comput. **37**, 48–59 (2015)
6. Shen, F., Shen, C., van den Hengel, A., Tang, Z.: Approximate least trimmed sum of squares fitting and applications in image analysis. IEEE Trans. Image Process. **22**, 1836–1847 (2013)
7. Huang, P., Gao, G.: Parameterless reconstructive discriminant analysis for feature extraction. Neurocomputing **190**, 50–59 (2016)
8. Xiong, C., Liu, L., Zhao, X., Yan, S., Kim, T.-K.: Convolutional fusion network for face verification in the wild. IEEE Trans. Circuits Syst. Video Technol. **26**, 517–528 (2016)
9. Gao, G., Yang, J., Jing, X.-Y., Shen, F., Yang, W., Yue, D.: Learning robust and discriminative low-rank representations for face recognition with occlusion. Pattern Recogn. **66**, 129–143 (2017)
10. Yang, J., Luo, L., Qian, J., Tai, Y., Zhang, F., Xu, Y.: Nuclear norm based matrix regression with applications to face recognition with occlusion and illumination changes. IEEE Trans. Pattern Anal. Mach. Intell. **39**, 156–171 (2017)
11. Chang, H., Yeung, D.-Y., Xiong, Y.: Super-resolution through neighbor embedding. In: CVPR, pp. 275–282. IEEE Press, Washington (2004)
12. Ma, X., Zhang, J., Qi, C.: Hallucinating face by position-patch. Pattern Recogn. **43**, 2224–2236 (2010)
13. Jung, C., Jiao, L., Liu, B., Gong, M.: Position-patch based face hallucination using convex optimization. IEEE Signal Process. Lett. **18**, 367–370 (2011)
14. Jiang, J., Hu, R., Wang, Z., Han, Z.: Noise robust face hallucination via locality-constrained representation. IEEE Trans. Multimedia **16**, 1268–1281 (2014)
15. Hennings-Yeomans, P.H., Baker, S., Kumar, B.V.: Simultaneous super-resolution and feature extraction for recognition of low-resolution faces. In: CVPR, pp. 1–8. IEEE Press, Anchorage (2008)
16. Biswas, S., Aggarwal, G., Flynn, P.J., Bowyer, K.W.: Pose-robust recognition of low-resolution face images. IEEE Trans. Pattern Anal. Mach. Intell. **35**, 3037–3049 (2013)
17. Jiang, J., Hu, R., Wang, Z., Cai, Z.: CDMMA: Coupled discriminant multi-manifold analysis for matching low-resolution face images. Sig. Process. **124**, 162–172 (2016)
18. Lu, T., Yang, W., Zhang, Y., Li, X., Xiong, Z.: Very low-resolution face recognition via semi-coupled locality-constrained representation. In: International Conference on Parallel and Distributed Systems (ICPADS), pp. 362–367 (2016)
19. Candes, E.J., Li, X.D., Ma, Y., Wright, J.: Robust principal component analysis? J. ACM **58**, 1–37 (2011)
20. Cai, J.F., Candes, E.J., Shen, Z.W.: A singular value thresholding algorithm for matrix completion. SIAM J. Optim. **20**, 1956–1982 (2010)
21. Martinez, A.M., Benavente, R.: The AR face database. CVC Technical report, vol. 24 (1998)
22. Wright, J., Yang, A.Y., Ganesh, A., Sastry, S.S., Ma, Y.: Robust face recognition via sparse representation. IEEE Trans. Pattern Anal. Mach. Intell. **31**, 210–227 (2009)

23. Lu, H., Li, B., Zhu, J., Li, Y., Li, Y., Xu, X., He, L., Li, X., Li, J., Serikawa, S.: Wound intensity correction and segmentation with convolutional neural networks. Concurrency Comput. Practice Exp. **29**(6), 1–8 (2017)
24. Lu, H., Li, Y., Chen, M., Kim, H., Serikawa, S.: Brain intelligence: go beyond artificial intelligence. (2017). arXiv:1706.01040

Face Recognition Benchmark with ID Photos

Dongshun Cui, Guanghao Zhang, Kai Hu, Wei Han
and Guang-Bin Huang

Abstract With the development of deep neural networks, researchers have developed lots of algorithms related to face and achieved comparable results to human-level performance on several databases. However, few feature extraction models work well in the real world when the subject which is to be recognized has limited samples, for example, only one ID photo can be obtained before the face recognition task. To our best knowledge, there is no face database which contains ID photos and pictures from the real world for a subject simultaneously. To fill this gap, we collected 100 celebrities' ID photos and their about 1000 stills or life pictures and formed a face database called **FDID**. Besides, we proposed a novel face recognition algorithm and evaluated it with this new database on the real-life videos.

Keywords Face recognition · Face recognition benchmark · ID photos · Total loss function · Real-life face recognition system

D. Cui (✉)
Energy Research Institute @ NTU (ERI@N), Interdisciplinary
Graduate School, Singapore, Singapore
e-mail: dcui002@ntu.edu.sg

D. Cui · G. Zhang · W. Han · G.-B. Huang
School of Electrical and Electronic Engineering,
Nanyang Technological University, Singapore, Singapore
e-mail: gzhang009@ntu.edu.sg

W. Han
e-mail: hanwei@ntu.edu.sg

G.-B. Huang
e-mail: egbhuang@ntu.edu.sg

K. Hu
College of Information Engineering, Xiangtan University, Xiangtan, China
e-mail: kaihu@xtu.edu.cn

© Springer International Publishing AG 2018
H. Lu and X. Xu (eds.), *Artificial Intelligence and Robotics*,
Studies in Computational Intelligence 752,
https://doi.org/10.1007/978-3-319-69877-9_4

1 Introduction

Tasks related to face like face recognition have been active research fields for many years. Human-level performance has been achieved by machine-learning based algorithms for face recognition on several existing datasets (*e.g.*, deep neural network [1–3] and extreme learning machines [4]). Some famous and popular face recognition database is summarized and compared in Table 1.

We consider the ID photo of a person is critical for face recognition since the only information we can obtain is ID photo under several specific scenarios before there is a request to recognize some person from the videos or in the real world. For example, it is likely that only ID photos are available before the arrest of criminals. This requires face recognition systems to extract enough useful features from an ID photo and recognize criminals with these representations. Another example is that when offering an intelligent service to some very important persons (*e.g.*, the head of a country) but it is inconvenient to scan them and acquire face samples. The reasons why we choose celebrities as the subjects are:

1. Large quantities. Huge data is the foundation of training a deep neural network [5–7], while the selected celebrities are very famous and large amounts of their stills and life photos (collectively called "non-ID photos") are available on the Internet.
2. Rich scenes. Compared to the ordinary people, celebrities live in much more rich scenes which leads to generating more face images with variable backgrounds and illumination conditions.

We have the largest average image quantity (donated as $\sharp$) for each person among all existing database. $\sharp$ is calculated by $\sharp = Q/N$, while Q is the total images of the database and N is the number of subjects. The comparison between the existing datasets and our FIFD dataset is shown in Table 1.

Table 1 The comparison between existing datasets

Dataset	Year[a]	N	Q	$\sharp$	ID photo	Pose variation
LFW	2007	5749	13233	2.30	N	Limited
WDRef	2012	2995	99773	33.31	N	Limited
CACD	2014	2000	163446	81.72	N	Limited
CASIA WebFace	2014	10575	494414	46.75	N	Full
FaceScrub	2014	530	106863	201.63	N	Limited
SFC	2014	4030	4400000	1091.81	N	Limited
CelebA	2015	10177	202599	19.91	N	Limited
IJB-A	2015	500	5712	11.42	N	Full
Megaface	2016	672057	1027060	1.49	N	Full
FIFD (Ours)	2017	100	112839	1128.39	Y	Full

[a]The time of publication of the corresponding papers

The main contributions of our work are:

- We have leveraged a face benchmark which consists of ID photos and non-ID photos of 100 (Chinese) celebrities for face recognition. The way to build the dataset and detailed analysis of the dataset are explained.
- We have proposed a novel architecture for general face recognition and tested it with the real-world videos to illustrate the quality of our FIFD dataset.

The rest of the paper is organized as follow: Sect. 2 introduces the related work on existing scientific methods on creating face databases and the state-of-the-art face recognition algorithms. Then we explain the details on how we collect our database and data diversity is shown by using the images from our database. In Section, we propose a novel face recognition architecture and show the results of our methods on the real-life videos by training a model with our database. A conclusion is given, and future work is claimed in Sect. 5.

2 Related Work

2.1 Protocol of Building Face Recognition Database

The database is essential for training face recognition algorithms to achieve models and evaluating their performance. And there is no doubt that algorithms benefit a lot from a comprehensive and exquisite database which requests a scientific process. We have introduced the size of the existing popular datasets in Sect. 1, and here the systematic procedure of creating a face recognition database is to be introduced.

The general steps of building a face database are collection, cleaning, and arrangement of face images. Experimental and real-life environment are the two sources for gathering images, and the Internet provides a convenient way to collect real world pictures. Face image cleaning consists of face detection, face alignment, and duplication elimination. All of these fields have been explored for over 20 years, but still provide no perfect solutions [8–10]. Finally, face image annotation (manually or automatically) and stored by the order are done.

2.2 Existing Face Recognition Algorithms

Here we list some state-of-the-art face recognition algorithms on LFW dataset. DeepFace is the first wide accepted face recognition algorithm that approaches human-level performance on LFW (97.35%) [1]. It follows the general process pipeline of face detection, face alignment, face representation and face verification. DeepFace trains an effective nine-layer deep neural network (DNN) and tune 120 million parameters with no weight sharing between local connections. Later,

DeepID, DeepID2, DeepID2+, and DeepID3 are proposed by modifying the structure of the network (for example, DeepID adopts Convolutional Network.) and increase the accuracy further [2, 11–13].

FaceNet is proposed by using triplet loss (each pair of triplet consists of an anchor sample x^a, a positive sample x^p, and a negative sample x^n) and increase the accuracy to 99.63%. The loss function is

$$\sum_i^N \left[\|x_i^a - x_i^p\|_2^2 + \alpha - \|x_i^a - x_i^n\|_2^2 \right]. \tag{1}$$

Here, α is a constant which guarantees the distance between positive and negative pairs.

3   Our Database

There are massive image data on the Internet, and we can find a huge amount of stills and life photos for most celebrities. In our task, only the celebrities of whom we can obtain a clear ID photo will be put into our dataset.

3.1   Collection Rules and Flowchart

The target of or work is to build a face recognition database which contains one ID photo and as more non-ID photos as possible. These non-ID photos should be diverse enough to simulate the real-world scenario. With the help of search engines (like Google, Baidu), we have obtained the ID photos of 100 Chinese celebrities so far. Folders are created and named after the celebrities, and ID photos are stored in the corresponding folders.

3.2   Resources of the ID and Non-ID Photos

There are lots of pictures on the Internet for the celebrities we can collect, and the first step is to find celebrities' ID photos. So far, we have gathered 100 celebrities' ID photos and stored these ID photos separately. With these celebrities' name and the full consideration of data diversities, we continue to search their stills and life photos mainly from the popular movie databases (*e.g.*, IMDb, Mtime), famous image search engines (*e.g.*, Google Images and Baidu Image), and some social networks (*e.g.*, Baidu Tieba, Sina Weibo).

Fig. 1 Some of the ID photos in our collected database. Note that these are Chinese ID-card photos, and people can have slight smiles when they take photos

3.3 Insight of the Proposed Database

We have created the face database (FDID) with the ID photos of the celebrities from the internet and as many of their corresponding non-ID photos as possible. We have randomly selected 36 celebrities (18 males and 18 females) and shown their ID photos in Fig. 1.

The reason why we pay more attention to the ID photo is that it usually provides one person's most information among all his face images. Here we have listed some properties of an ID photo in the passport [14]:

1. No head pose.
2. Neutral expression.
3. No occlusions (*e.g.*, glasses and marks) on the face.

Obviously, these rules echo the main challenges of real-world face recognition applications since non-ID photos always don't fulfill one or several of these rules. To make an intuitive comparison, we randomly select four celebrities (two males and two females) to show the diversity contents of our database. Examples of face images with large-angle head pose, various expressions, and occlusions from our database are shown in Fig. 2a, b, c respectively.

(a) Different head poses.   (b) Various facial expressions.

(c) Occlusions.

Fig. 2 Examples of face images from our database

4 Proposed Method and Results

Based on our database including ID photos and non-ID photos, we re-design the architecture of the face recognition as shown in Fig. 3.

In this section, we will introduce a novel and general face recognition loss function for our database. We point two kinds of intra-person variances based on the assumptions of the same person's non-ID photo should be similar to his/her ID photo and the same person should look similar to his/her other non-ID photo.

We use x_{ij} indicates the j-th photo of the i-th person in the dataset. To distinguish the ID photo and non-ID photos, we assume $j = 0$ denotes the former while $j \in [1, m]$ indexes the latter. Assume $r(x)$ as the representation of sample x, it can be manually descriptors (*e.g.*, LBP, Gabor, and Eigenvector) or learned by a neural network (*e.g.*, AlexNet, PCANet, and CNN). Dissimilarity function is expressed as $d(\cdot)$. These two kinds of intra-person variances yield the following two loss functions.

The first intra-person loss function is:

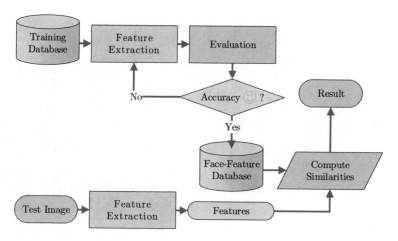

Fig. 3 The proposed architecture of the face recognition

$$\mathcal{L}_1 = \sum_{i\in[1,n]}\sum_{j\in[1,m]} d_1(r(x_{i0}), r(x_{ij})) \tag{2}$$

The second intra-person loss function is:

$$\mathcal{L}_2 = \sum_{i\in[1,n]}\sum_{j_1,j_2\in[1,m]} d_2(r(x_{ij_1}), r(x_{ij_2})). \tag{3}$$

Intra-person loss functions constraint the similarity between the photos of the same person, and for face recognition, we also need to constraint the dissimilarity between the photos of the different persons.

The inter-person loss function is:

$$\mathcal{L}_3 = \sum_{i_1,i_2\in[1,n]}\sum_{j_1,j_2\in[1,m]} d_3(r(x_{i_1j_1}), r(x_{i_2j_2})). \tag{4}$$

So, the total loss function $\mathcal{L}_{total}$ for the set of face images is:

$$\mathcal{L}_{total} = \lambda_1\mathcal{L}_1 + \lambda_2\mathcal{L}_2 - \lambda_3\mathcal{L}_3 \tag{5}$$

where $\lambda_1, \lambda_2, \lambda_3$ are the weights of $\mathcal{L}_1, \mathcal{L}_2, \mathcal{L}_3$ respectively. Our target is to minimize this loss function by learning an effective model to represent the face images, which is expressed as:

$$\mathcal{T}(r, \lambda_i) = min(\mathcal{L}_{total}). \tag{6}$$

Currently, most face recognition models are the instances of Eq. 6 by setting $\lambda_1 = \lambda_2$.

Fig. 4 Results of the proposed method on real-life videos (frame sequences) by training on our database. Each row shows the face recognition result of the same celebrity

We train a model with our dataset by adopting a similar network structure proposed in [1] by replacing the softmax loss function with ours and test it with our database and some real-life videos. ID and non-ID photos of 80 of our collected celebrities are randomly selected for training and ten celebrities' photos for verifying, and the best feature model (mainly tuning the parameters of the networks) and dissimilarity computation algorithm are achieved. The remaining 10 celebrities' non-ID photos are used to test, and we achieve an accuracy of near 80%. Besides, we collect these 10 celebrities' videos from YouTube and transfer them into frames. Then we implement the same procedures of preprocessing (*e.g.*, face detection), and extract features with the optimized model. Finally, we compute the dissimilarity between features of the input face images and all the candidate ID photos and output the celebrity's name which corresponded to the minimum dissimilarity value. We have an ID photo and extract features from it. Then for the input videos, we do the face detection, feature extraction, and dissimilarity computation. Finally, we output the face identification results which are demonstrated in Fig. 4.

5 Conclusion

We have created a new database which contains ID photos of 100 celebrities with their over 1000 stills and life photos. Details on how to build collect this database are introduced, and the data diversity on faces with different head poses, facial expressions, occlusions, illuminations, degrees of toilette, and ages are shown. Besides, we proposed a novel architecture on face recognition when the ID photos are available, and new total loss function which contains two intra-person loss functions and one inter-person loss function are presented. Models are trained with the proposed database using our target function, and video-level experiments are performed to demonstrate the meaning of our database and the effectiveness of the proposed method.

References

1. Taigman, Y., Yang, M., Ranzato, M., Wolf, L.: Deepface: closing the gap to human-level performance in face verification. In: Proceedings of the IEEE Conference on Computer Vision and Pattern Recognition, pp. 1701–1708 (2014)
2. Sun, Y., Wang, X., Tang, X.: Deeply learned face representations are sparse, selective, and robust. In Proceedings of the IEEE Conference on Computer Vision and Pattern Recognition, pp. 2892–2900 (2015)
3. Schroff, F., Kalenichenko, D., Philbin, J.: Facenet: a unified embedding for face recognition and clustering. In: Proceedings of the IEEE Conference on Computer Vision and Pattern Recognition, pp. 815–823 (2015)
4. Tang, J., Deng, C., Huang, G.-B.: Extreme learning machine for multilayer perceptron. IEEE trans. Neural Netw. Learn. Syst. **27**(4), 809–821 (2016)
5. Lu, H., Zhang, L., Serikawa, S.: Maximum local energy: an effective approach for multisensor image fusion in beyond wavelet transform domain. Comput. Math. Appl. **64**(5), 996–1003 (2012)
6. Xu, X., He, L., Shimada, A., Taniguchi, R.-I., Lu, H.: Learning unified binary codes for cross-modal retrieval via latent semantic hashing. Neurocomputing **213**, 191–203 (2016)
7. Lu, H., Li, B., Zhu, J., Li, Y., Li, Y., Xu, X., He, L., Li, X., Li, J., Serikawa, S.: Wound intensity correction and segmentation with convolutional neural networks. Concurr. Comput. Pract. Exp. **29**(6) (2017)
8. Kawulok, M., Celebi, M.E., Smolka, B.: Advances in Face Detection and Facial Image Analysis. Springer (2016)
9. Liu, Q., Deng, J., Tao, D.: Dual sparse constrained cascade regression for robust face alignment. IEEE Trans. Image Process. **25**(2), 700–712 (2016)
10. Tang, J., Li, Z., Wang, M., Zhao, R.: Neighborhood discriminant hashing for large-scale image retrieval. IEEE Trans Image Process. **24**(9), 2827–2840 (2015)
11. Sun, Y., Wang, X., Tang, X.: Deep learning face representation from predicting 10,000 classes. In: Proceedings of the IEEE Conference on Computer Vision and Pattern Recognition, pp. 1891–1898 (2014)
12. Sun, Y., Chen, Y., Wang, X., Tang, X.: Deep learning face representation by joint identification-verification. In: Advances in Neural Information Processing Systems, pp. 1988–1996 (2014)
13. Sun, Y., Liang, D., Wang, X., Tang, X.: Deepid3: face recognition with very deep neural networks (2015). arXiv:1502.00873
14. Rules for passport photos. https://www.gov.uk/photos-for-passports/photo-requirements. Accessed 27 Feb 2017

Scene Relighting Using a Single Reference Image Through Material Constrained Layer Decomposition

Xin Jin, Yannan Li, Ningning Liu, Xiaodong Li, Quan Zhou, Yulu Tian and Shiming Ge

Abstract Image relighting is to change the illumination of an image to a target illumination effect without known the original scene geometry, material information and illumination condition. We propose a novel outdoor scene relighting method, which needs only a single reference image and is based on material constrained layer decomposition. Firstly, the material map is extracted from the input image. Then, the reference image is warped to the input image through patch match based image warping. Lastly, the input image is relighted using material constrained layer decomposition. The experimental results reveal that our method can produce similar illumination effect as that of the reference image on the input image using only a single reference image.

Keywords Image relighting · Single reference image · Material map
Layer decomposition

X. Jin · Y. Li · X. Li (✉) · Y. Tian
Beijing Electronic Science and Technology Institute, Beijing 100070,
People's Republic of China
e-mail: lxd@besti.edu.cn

N. Liu
University of International Business and Economics, Beijing 100029,
People's Republic of China

Q. Zhou
National Engineering Research Center of Communications and Networking,
Nanjing University of Posts and Telecommunications, Nanjing,
People's Republic of China

Q. Zhou
State Key Laboratory for Novel Software Technology, Nanjing University,
Nanjing, People's Republic of China

S. Ge (✉)
Institute of Information Engineering, Chinese Academy of Sciences, Beijing 100093,
People's Republic of China
e-mail: geshiming@iie.ac.cn

© Springer International Publishing AG 2018
H. Lu and X. Xu (eds.), *Artificial Intelligence and Robotics*,
Studies in Computational Intelligence 752,
https://doi.org/10.1007/978-3-319-69877-9_5

1   Introduction

Image relighting is a hot topic in the communities of computer vision, image processing and computational photography. The applications of image relighting include visual communication, film production and digital entertainment, etc.

Image relighting is to change the illumination of an image to a target illumination effect without known the original scene geometry, material information and illumination condition. Comparing with face, object and indoor scene, two main challenges are for outdoor scene relighting: (1) large scale outdoor scene with multiple objects, which are not easy to reconstruct; (2) complex illumination in outdoor scene, which is hard to be controlled manually. Recently, reference image based image relighting has shown great potential [1–4]. Currently, for face relighting, the reference images are changed from multiple and a pair [5] to a single [3]. For object relighting [2, 6] and scene relighting [1], multiple or a pair reference images are still needed [7–11], [16–18].

We propose a novel outdoor scene relighting method, which needs only a single reference image and is based on material constrained layer decomposition. Firstly, the material map is extracted from the input image. Then, the reference image is warped to the input image through patch match based image warping. Lastly, the input image is relighted using material constrained layer decomposition. The experimental results reveal that our method can produce similar illumination effect as that of the reference image on the input image using only a single reference image.

2   Scene Relighting

2.1   Method Overview

Our proposed method can be divided into 4 steps, as shown in Fig. 1: (1) the input image is segmented to the material map using the method of Bell et al. [12]. Every

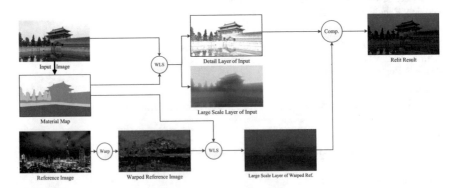

Fig. 1 Scene relighting using only a single reference image

pixel of the material map is assigned by a material label; (2) the reference image is warped to the structure of the input image by the patch match warping; (3) each channel of the input image and the reference is decomposed to large scale layer and detail layer under material constrain; (4) the final relit results are obtained by composing the details of the input image and the large scale of the warped reference image.

2.2 Material Segmentation

The input image is segmented according to the material of each pixel. We use the method of Bell et al. [12] to obtain material label of each pixel. We make the material segmentation because that in different material region, different relighting operations should be conducted. We select 9 sorts of materials, which often appear in outdoor scene images, as shown in Fig. 2. We recolor each pixel according to the material label to get the material map. The first and the third lines are the input images. The second and the forth lines are the corresponding material maps.

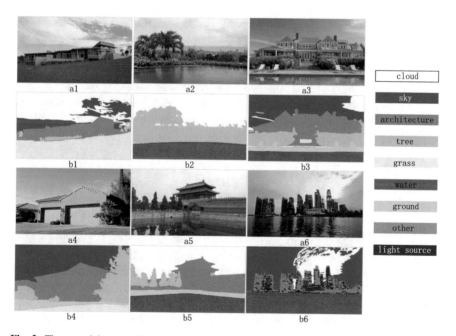

Fig. 2 The material maps of the input images

2.3 Reference Image Warping

In face image relighting, the reference face image can be warped by face landmark detection/face alignment. However, in outdoor scene, we cannot find such similar structure easily. The outdoor scene contains multiple objects. Thus, we use the patch match method to warp the reference image to the input image, i.e. to align the reference and the input image. The patch match algorithm is similar as the method of Barnes et al. [13]. We use the neighbor patches whose best matched patches have already been found to improve matching result of current patch. The difference from Barnes et al. [13] is that we use 4 neighbor patches instead of 3 ones.

2.4 Layer Decomposition and Composition

We use the WLS filter [14] to decompose image into large scale layer and detail layer, which can be considered as the illumination component and non-illumination component. Using the large scale layer of the warped reference to substitute the large scale layer of the input can produce the final relit result. The outdoor scene contains various objects with various materials. Thus for different material, different decomposition parameters should be used. Each channel l of the input image and the reference image is filtered to a large scale layer s. The detail layer d is obtained by: $d = l/s$.

The original WLS filter uses the same smoothness level over the whole image. When using the WLS filter for our scene relighting task, we need make regions with different materials with different smooth levels. Thus, we set different smoothness levels in regions with different materials. We modified the original WLS [14] as:

$$E = |l - s|^2 + H(\nabla_s, \nabla l) \tag{1}$$

$$H(\nabla_s, \nabla l) = \sum_p \left(\lambda(p) \left(\frac{(\partial s/\partial x)_p^2}{(\partial l/\partial x)_p^\alpha + \varepsilon} + \frac{(\partial s/\partial y)_p^2}{(\partial l/\partial y)_p^\alpha + \varepsilon} \right) \right), \tag{2}$$

where, $|l - s|^2$ is the data term, which is to let l and s as similar as possible, i.e., to minimize the distance between l and s. $H(\nabla_s, \nabla l)$ is the regularization (smoothness) term, which makes s as smooth as possible, i.e. to minimize the partial derivative of s. p is the pixel of the image. α controls over the affinities by non-linearly scaling the gradients. Increasing α will result in sharper preserved edges. λ is the balance factor between the data term and the smoothness term. Increasing λ will produce smoother images. ε is a very small number, so as to avoid the division by 0. Our λ is the smoothness level constrained by different materials, using the material map derived in Sect. 3.2:

$$\lambda = \nabla 1 + gray(l_m)/255, \tag{3}$$

where, $\nabla 1$ is the gradient of l. l_m is the material map of l, and the gray is the gray value of l_m:

$$gray = (R * 0.2989 + G * 0.587 + B * 0.114). \tag{4}$$

The minimization of Eqs. (1) and (2) can be solved by the off-the-shell methods such as Lischinski [15]. At last, using the large scale layer of the warped reference to substitute the large scale layer of the input can produce the final relit result.

3 Experimental Results

In this section, we show the experimental results of our proposed method and the comparison with the state of the art method.

3.1 The Scene Relit Results

The relit results of our method are shown in Fig. 3. (a): multiple input images, (b): the same reference image, (c): warped reference image, (d): relit results of input images using (c). The experimental results reveal that, the relit input image have similar illumination effect as that of the reference image.

Fig. 3 The relit results of our method

Fig. 4 Comparison with [1]

3.2 Comparison with Other Methods

We compare our method with the state of the art method [1], which needs a time-lapse video captured by a fixed camera working for 24 h. While our method needs only a single reference image. We randomly select 5 input images for comparison. As shown in Fig. 4, (a): multiple input images, (b) the reference image, (c): warped reference images to corresponding input images, (d): warped reference images using method of [1], note that they need a time-lapse video for warping, (e): the relit results using our proposed method, (f): the relit results using the method of [1]. The results reveal that our method can produce similar relit results as those of [1], with only a single reference image.

4 Conclusion

We propose a novel outdoor scene relighting method, which needs only a single reference image and is based on material constrained layer decomposition. The experimental results reveal that our method can produce similar illumination effect as that of the reference image on the input image using only a single reference image.

Acknowledgements We thank all the reviewers and PCs. This work is partially supported by the National Natural Science Foundation of China (Grant NO.61402021, 61401228, 61640216), the Science and Technology Project of the State Archives Administrator (Grant NO. 2015-B-10), the open funding project of State Key Laboratory of Virtual Reality Technology and Systems, Beihang University (Grant NO. BUAA-VR-16KF-09), the Fundamental Research Funds for the Central Universities (Grant NO.2016LG03, 2016LG04), the China Postdoctoral Science Foundation (Grant NO.2015M581841), and the Postdoctoral Science Foundation of Jiangsu Province (Grant NO.1501019A).

References

1. Shih, Y., Paris, S., Durand, F., Freeman, W.T.: Data-driven hallucination of different times of day from a single outdoor photo. ACM Trans. Graph. **32**(6), 11, Article 200 (2013)
2. Haber, T., Fuchs, C., Bekaer, P., et al.: Relighting objects from image collections. In: IEEE Conference on Computer Vision and Pattern Recognition, 2009 (CVPR 2009), pp. 627–634. IEEE (2009)
3. Chen, X., Wu, H., Jin, X., Zhao, Q.: Face illumination manipulation using a single reference image by adaptive layer decomposition. IEEE Trans. Image Process. (TIP) **22**(11), 4249–4259 (2013)
4. Peers, P., Tamura, N., Matusik, W., Debevec, P.: Post-production facial performance relighting using reflectance transfer. ACM Trans. Graph. **26**(3), Article 52 (2007)
5. Chen, J., Su, G., He, J., Ben, S.: Face image relighting using locally constrained global optimization. Computer Vision—ECCV 2010 Volume 6314 of the series. Lecture Notes in Computer Science, pp. 44–57 (2010)
6. Jin, X., Tian, Y., Ye, C., Chi, J., Li, X., Zhao, G.: Object image relighting through patch match warping and color transfer. In: The 16th International Conference on Virtual Reality and Visualization (ICVRV), Hangzhou, China, pp. 235–241, 23–25 Sept 2016
7. Lu, H., Guna, J., Dansereau, D.G.: Introduction to the special section on artificial intelligence and computer vision. Comput. Electr. Eng. **58**, 444–446 (2017)
8. Lu, H., Li, Y., Nakashima, S., Kim, H., Serikawa, S.: Underwater image super-resolution by descattering and fusion. IEEE Access **5**, 670–679 (2017)
9. Lu, H., Li, Y., Zhang, L., Serikawa, S.: Contrast enhancement for images in turbid water. J. Opt. Soc. Am. A. **32**(5), 886–893 (2015)
10. Lu, H., Li, Y., Nakashima, S., Serikawa, S.: Turbidity underwater image restoration using spectral properties and light compensation. IEICE Trans. Inf. Syst. **99**(1), 219–227 (2016)
11. Zhou, Q., Zheng, B., Zhu, W., Latecki, L.J.: Multi-scale context for scene labeling via flexible segmentation graph. Pattern Recogn. **2016**(59), 312–324 (2016)
12. Bell, S., Upchurch, P., Snavely, N., Bala, K.: Material recognition in the wild with the materials in context database. In: IEEE Conference on Computer Vision and Pattern Recognition, (CVPR) 2015, Boston, MA, USA, 7–12 June 2015
13. Barnes, C., Shechtman, E., Goldman, D., Finkelstein, A.: The generalized patchmatch correspondence algorithm. Computer Vision–ECCV 2010, 29–43 (2010)
14. Farbman, Z., Fattal, R., Lischinski, D., Szeliski, R.: Edgepreserving decompositions for multi-scale tone and detail manipulation. ACM Trans. Graph. (Proceedings of ACM SIGGRAPH 2008), **27**(3) (2008)
15. Lischinski, D., Farbman, Z., Uyttendaele, M., Szeliski, R.: Interactive local adjustment of tonal values. ACM Trans. Graph. **25**(3): 646–653 (2006)

16. Wang, Y., Liu, Z., Hua, G., Wen, Z., Zhang, Z., Samaras, D.: Face re-lighting from a single image under harsh lighting conditions. In: Proc. CVPR (2007)
17. Eisemann, E., Durand, F.: Flash photography enhancement via intrinsic relighting. ACM Trans. Graph. **23(3)**, 673-678 (2004)
18. Agrawal, A., Raskar, R.: Gradient domain manipulation techniques in vision and graphics. In: ICCV 2007 Courses (2007)

Applying Adaptive Actor-Critic Learning to Human Upper Lime Lifting Motion

Ting Wang and Ryad Chellali

Abstract An adaptive reinforcement learning method designed to facilitate the on-line lifting motion of the human forearm is here proposed. Its purpose is to use the control based on the proposed learning method to perform the lifting motion. The learning algorithm is an actor-critic learning based on the neural network that used the normalized radial basis function. The paper shows a simulation of the motion of the forearm lifting process. As shown in the results, the forearm continues to lift from a horizontal position to a vertical position. During this process, both the state space and action space are continuous.

Keywords Reinforcement learning · Normalized radial basis function Neural network · Actor-critic

1 Introduction

With the development of robotics research, the requirements for robot learning ability are increasing so they may perform more complex motor behavior. For complicated motions, such as striking a target, it is impossible to achieve motion control using traditional algorithms. The robot instead needs to have the ability to imitate human behavior, in order to attain the intelligence necessary for this task. It can achieve self-compensation correction and interact with random dynamic environments [1].

The main advantage of robot learning is to find an effective strategy to control complex motion tasks in the case that the conventional methods do not work. Imitation learning is a kind of robot learning, also known as learning from

T. Wang (✉) · R. Chellali
College of Electrical Engineering and Control Science, Nanjing Tech University,
30, Puzhu Sourth Road, Pukou District, Nanjing 211800, Jiangsu Province, China
e-mail: wangting0310@njtech.edu.cn

R. Chellali
e-mail: rchellali@njtech.edu.cn

© Springer International Publishing AG 2018
H. Lu and X. Xu (eds.), *Artificial Intelligence and Robotics*,
Studies in Computational Intelligence 752,
https://doi.org/10.1007/978-3-319-69877-9_6

45

programming by demonstration [2–4]. It plays an important role in bionic robotics research, and is a significant embodiment of robot intelligence. Imitation learning has been used for many intricate bionic motion tasks, such as batting, drumming, walking, performing billiards, pancake-flipping, catching ball-in-a-cup, and similar actions [5–13].

In this project, we studied the imitation of basic lifting motion of human forearm by actor-critic learning. This is the beginning of our research into imitation learning in humanoid robots. Through the actor-critic learning control based on the normalized radial basis function (NRBF) neural network, the simulation results validate the algorithm's ability to facilitate the lifting motion. For the next step, the research will focus on using humanoid robots to accomplish more complex motion tasks through actor-critic learning control.

The rest of paper is organized as follows: Sect. 2 introduces the state of art of planar human arm model and NRBF neural network, Sect. 3 demonstrated results of simulation of lifting motion by human forearm, and some conclusions are described in the conclusion section.

2   State of the Art

2.1   Dynamic Arm Simulator Model

The biomechanical dynamic arm simulator (DAS) model is illustrated in Fig. 1, and a detailed description is given in a previous work [14]. The detailed biomechanical simulation of a human arm under functional electrical stimulation can be found in a series of previous works [13–20]. The model is regarded as an accurate model of an ideal arm. Assuming that there is no friction during the motion, the arm involves two joints (the shoulder and elbow), driven by six muscles. A three-element Hill model was used for each muscle and the muscle was simulated using two differential equations. One differential equation is used to describe the activation, the

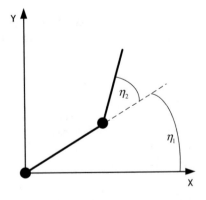

Fig. 1   The DAS constituted two joints and six muscles. Antagonistic muscle pairs are as follows: monoarticular shoulder muscles, monoarticular elbow muscles, and biarticular muscles. The detailed model and its explanation can be found in a previous work [14]

other to describe the contraction [21]. Because the internal muscle states were not dominant over the controller, the muscle force was indirectly controlled by muscle dynamics [14, 21].

The DAS model has four state variables as: $s_t = (\eta_1, \eta_2, \dot{\eta}_1, \dot{\eta}_2)$, which are the rotation angles and angle velocities of two joints, respectively. It is here assumed that η_1^d, η_2^d are the desired joint angles, and the desired joint angle velocities are zero. The goal is to move the forearm from its random initial state to a randomly chosen stationary target during a given period. As an example, we choose to vary η_2 from $0°$ to $90°$ and vary η_1 from $0°$ to $45°$ at the same time.

3 NRBF Network

3.1 Actor-Critic Learning Based on the NRBF Network

The structure of the actor-critic learning based on NRBF network is illustrated in Fig. 2. It is an adaptive reinforcement learning structure, which used NRBF network to approximate critic function and actor function. Actor function projected the perception state to a random policy of action probability. Critic functions affect the evaluation of the current policy. Actors and critics can be established through the NRBF network. It not only reduces the forward calculation time, but also responds to the dynamic variation of the unknown system. The network can be adjusted adaptively so that it has plasticity.

The actor and critic share the common input layer and latent layer, shown in Fig. 3. In the actor-critic learning system, the actor acts as the controller, projecting from state space $s_t = [s_{t1}, s_{t2}, \ldots, s_{tn}]^T \in \mathbf{R}^n$ to actions space $A(s_t) \in \mathbf{R}^m$, where n

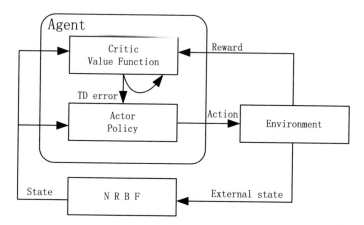

Fig. 2 Actor-Critic learning structure based on NRBF network

Fig. 3 Actor-Critic learning
structure based on NRBF
network

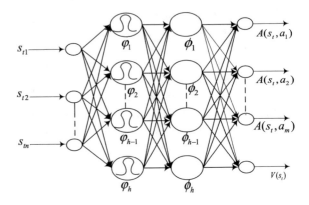

and m correspond to dimensions of state space and action space. At the same time,
the evaluation function acquired from current policy learning is defined as follows:

$$V(s_t) = E[\sum_{k=0}^{\infty} \gamma^k r_{t+k}] \in R^l. \tag{1}$$

In Eq. (1), γ is discount coefficient, and $0 < \gamma < 1$, which is used to determine
the ratio between the delay rewards and immediate rewards.

The output of the latent layer of NRBF network adopted Gaussian function.
When the input state is s_t, the output of the j-th latent layer $\varphi_j(s_t)$, and its corresponding normalized output $\phi_j(s_t)$ can be expressed as follows:

$$\varphi_j(s_t) = \exp\left[-\frac{(s_t - \mu_j)^2}{2\sigma_j^2}\right] \tag{2}$$

$$\phi_j(s_t) = \frac{\varphi_j(s_t)}{\sum_{t=1}^{h} \varphi(s_t)}. \tag{3}$$

In Eqs. (2) and (3), μ_j, σ_j are the center and the width of the j-th joint of the latent
layer, respectively. h is the number of joints in the j-th latent layer.

The actor network output $A_k(s_t)$ and critic network output $V(s_t)$ are written as
follows:

$$A_k(s_t) = \sum_{j=1}^{h} w_{kj} \varphi_j(s_t) \tag{4}$$

$$V(s_t) = \sum_{j=1}^{h} v_j \varphi_j(s_t) \tag{5}$$

Here, w_{kj} is the weight from the j-th joint of the latent layer to the k-th output
joint of the actor network.

Here, the output $A_k(s_t)$ is not calculated directly in the system, while it is firstly sent to stochastic action modify (SAM). SAM is equivalent to a search mechanism that acts on the actual control of the system. The actual control variable $\hat{A}_k(s_t)$ is calculated as follows. It is determined by Gaussian distribution.

$$p\big(\hat{A}_k(s_t)\big) = \exp\left[-\frac{\big(\hat{A}_k(s_t) - A_k(s_t)\big)^2}{\rho_t^2} \right] \tag{6}$$

$$\rho_t = \frac{k_1}{1 + \exp(k_2 V(s_t))}. \tag{7}$$

In Eqs. (6) and (7), ρ_t is a monotonous decrease function. Larger $V(s_t)$ corresponds to smaller ρ_t, so as to make the actual system output approach to the recommend output.

3.2 Updating the Actor and Critic Networks

In this paper, the temporal difference (TD) method is adopted to learn the actor and critic network. The TD error δ_t is calculated as follows:

$$\delta_t = r_t + \gamma V(s_{t+1}) - V(s_t) \tag{8}$$

Here, s_t, s_{t+1} are the states of time t and time t + 1, respectively. r_t is the rewards function.

The approximate policy gradient estimation algorithm is used to update the actor network. The weight adjustment w_{kj} between the j-th latent layer and k-th output layer is:

$$\Delta w_{kj} = \alpha_A \delta_t \left[\frac{\hat{A}_k - A_k}{\rho}\right]_{t-1} \varphi_j(t-1) \tag{9}$$

Here, α_A is the learning rate of actor network.

The center and width adjustments of the actor network latent layer are expressed as follows, where $\alpha_\mu, \alpha_\sigma$ are the learning rates of center and width, respectively.

$$\Delta \mu_j = \alpha_\mu \delta_t \left[\frac{\hat{A}_k - A_k}{\rho}\right]_{t-1} \left[\frac{\varphi_j(1-\varphi_j)w_{kj}(s-\mu_j)}{\sigma_j^2}\right]_{t-1} \tag{10}$$

$$\Delta \sigma_j = \alpha_\mu \delta_t \left[\frac{\hat{A}_k - A_k}{\rho}\right]_{t-1} \left[\frac{\varphi_j(1-\varphi_j)w_{kj}(s-\mu_j)}{\sigma_j^3}\right]_{t-1} \tag{11}$$

As in the actor network, the TD (λ) (λ is the attenuation coefficient, and $0 < \lambda < 1$) is also used to update the critic network. The weight adjustment of the critic network is as follows:

$$\Delta v_j = \alpha_C \delta_t \sum_{k=1}^{t-1} \lambda^{t-k-1} \nabla_{v_j} V_k \qquad (12)$$

where α_C is the learning rate of Critic network. $\sum_{k=1}^{t-1} \lambda^{t-k-1} \nabla_{v_j} V_k$ is defined as Eligibility trace e_t, that is, $e_t = \sum_{k=1}^{t-1} \lambda^{t-k-1} \nabla_{v_j} V_k$. Transfer e_t to recursive form as follows:

$$e_t = \nabla_{v_j} V_{k-1} + \lambda \sum_{k=1}^{t-2} \lambda^{t-k-1} \nabla_{v_j} V_k = \varphi_j(t-1) + \lambda e_{t-1} \qquad (13)$$

Substitute Eq. (13) from Eq. (12), the weight adjustment v_j is as follows:

$$\Delta v_j = \alpha_C \delta_t (\varphi_t(t-1) + \lambda e_{t-1}). \qquad (14)$$

4   Simulation and Results

The results of the simulation performed using the learning algorithm illustrated in Sect. 2, are displayed in Fig. 4. Parameters are setting as: $\alpha_C = 0.2, \alpha_A = 0.2, \gamma = 0.8, \lambda = 0.85, \alpha_\mu = 0.25, \alpha_\sigma = 0.25$. The rewards function is identical to that used by Jagodnik and van den Bogertto prevent joint angle error and high muscle stimulation [14]. State variables of forearm joint angles are: $s_t = (\eta_1, \eta_2, \dot{\eta}_1, \dot{\eta}_2)$, the desired state $\eta_{d1} = 90°, \eta_{d2} = 45°$. Here, sacrificing the accuracy of biology, assume that there are only two muscles in the forearm and that each muscle has 5 control elements. Actions can be stimulated using different interactions among muscle control elements, which can be computed as follows:

$$a_i = \frac{1}{2}(a_{i1} cos\eta_1 + a_{i2} cos\eta_2 + a_{i3} cos(\eta_1 + \eta_2) + a_{i4}\eta_1 + a_{i5}\eta_5) \, i \in \{1, 2\} \qquad (15)$$

It should be noted that Eq. (15) $a_i \in [0, 1]$ must use at least two muscle elements per muscle to allow the muscle to achieve maximum stimulation.

Actions are chosen as in a previous work [14]. After the 100 run, the learning system can keep the system status in 8000 iterations within the set range. In this way, the ability to balance can be successfully achieved. Through the actor-critic learning mechanism and the LQR controller, the whole arm arrives in the desired position. That is, η_1 and η_2 finally reached η_1^d and η_2^d. The results of learning are illustrated in Fig. 4a. Control results of angles (η_1, η_2) and angle velocities ($\dot{\eta}_1$, $\dot{\eta}_2$)

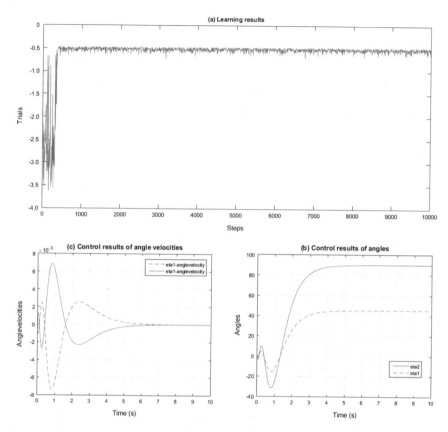

Fig. 4 Results of simulation

are displayed in Fig. 4b and 4c, respectively. From the results, based on the NRBF neural network, the learning rate is rapid, requiring relatively little computation and showing good convergence. The control showed high accuracy and strong stability.

5 Conclusion

The paper proposed a humanoid robot imitation learning algorithm and validated by the simulation. The learning algorithm is an actor-critic algorithm based on the NRBF network. It is designed to implement the lifting process of the human forearm. From the result of simulation, the simple human lifting motion has been achieved. Future work will focus on completing the imitation learning process through control of real humanoid robots with motors. The work will also focus on accomplishing more complicated human forearm motions.

References

1. Schaal, S., Atkeson, C.: Learning control in robotics. IEEE Robot. Autom. Mag. **17**(2), 20–29 (2010)
2. Schaal, S.: Is imitation learning the route to humanoid robots? Trends Cogn. Sci. **3**(6), 233–242 (1999)
3. Argall, B.D., Chernova, S., Veloso, M.: A survey of robot learning from demonstration. Robot. Auton. Syst. **57**(5), 469–483 (2009)
4. Billard, B.D., Calinon, S.: Robotic programming by demonstration. Springer Handbook of Robotics, pp. 1371–1394. Springer, Berlin (2008)
5. Mulling, K., Kober, J., Kroemer, O.: Learning to select and generalize striking movements in robot table tennis. Int. J. Robot. Res. **32**(3), 263–279 (2013)
6. Pastor, P., Kalakrishnan, M., Chitta, S.: Skill learning and task outcome prediction for manipulation. In: IEEE International Conference on Robotics and Automation, pp. 3828–3824. IEEE, Shanghai (2011)
7. Kormushev, P., Calinon, S., Caldwell, D.G.: Robot motor skill coordination with EM-based reinforcement learning. In: IEEE/RSJ International Conference on Intelligent Robots and Systems, pp. 3232–3237. IEEE, Taipei (2010)
8. Kober, J., Peters, J.: Imitation and reinforcement learning. Robot. Autom. Mag. **17**(2), 55–62 (2010)
9. Kormushev, P., Calinon, S., Caldwell, D.G.: Imitation learning of positional and force skills demonstrated via kinesthetic teaching and haptic input. Adv. Robot. **25**(5), 581–603 (2011)
10. Abbeel, P., Coates A.: Autonomous helicopter aerobatics through apprentices ship learning. Int. J. Robot. Res. 1–31 (2010)
11. Lee, K., Su, Y., Kim, T.K.: A syntactic approach to robot imitation learning using probabilistic activity grammars. Robot. Autom. Syst. **61**(12), 1323–1334 (2013)
12. Blana, D., Kirsch, R.F., Chadwick, E.K.: Combined feed forward and feedback control of a redundant, nonlinear, dynamic musculoskeletal system. Med. Biol. Eng. Comput. **47**, 533–542 (2009)
13. Jagodnik, K., van den Bogert, A.: A proportional derivative FES controller for planar arm movement. In: 12th Annual Conference International FES Society, Philadelphia, PA (2007)
14. Thomas, P., Branicky, M., van den Bogert, A., Jagodnik, K.: Application of the actor-critic architecture to functional electrical stimulation control of a human arm. In: Proceeding Innovations Applied Artificial Intelligence Conference, pp. 165–172 (2009)
15. Philip, S.T., Andrew, G.B.: Motor primitive discovery. In: Proceedings of the IEEE Conference on Development and Learning and Epigenetic Robotics (2012)
16. Lu, H., Li, Y., Uemura, T., Ge, Z., Xu, X.: FDCNet: Filtering deep convolution network for marine organism classification. In: Multimedia Tools and Applications, pp. 1–14 (2017)
17. He, L., Xing, X., Humin, L., Yang, Y., Fumin, S., Yuxuan, Z., Xing, X., Tao, H.: Unsupervised cross-model retrieval through adversarial learning. In: Proceeding of 2017 IEEE International Conference on Multimedia and Expo (2017)
18. Xu, X., He, L., Shimada, A., Taniguchi, R.I., Lu, H.: Learning unified binary codes for cross-modal retrieval via latent semantic hashing. In: Neurocomputing, **213**, 191–203 (2016)
19. Humin, L., Yujie, L., Min, Chen., Hyoungseop, Kim., Serikawa, S.: Brain intelligence: go beyond artificial intelligence. In: Mobile Networks Application, pp. 1–10 (2017)
20. Serikawa, S., Lu, H.: Underwater image dehazing using joint trilateral filter. Comput. Electr. Eng. **40**(1), 41–50 (2014)
21. McLean, S.G., Su, A., van den Bogert, A.J.: Development and validation of a 3-D model to predict knee joint loading during dynamic movement. J. Biomech. Eng. **125**(6), 864–874 (2003)

A Demand-Based Allocation Mechanism for Virtual Machine

Ling Teng, Hejing Geng, Zhou Yang and Junwu Zhu

Abstract In the Iaas service mode for cloud computing, cloud providers allocate resources in the form of Virtual Machines (VM) to cloud users via auction mechanism. The existing auction mechanism lacks self-adapting adjustment to market changes. An improved online auction mechanism by taking into account the changes in demand during peak and trough period in the allocation scheme has been proposed, so that the auctioneer can make decisions reasonably, improve resource utilization rate, and bring higher profits. Firstly, we present an auction framework for VM allocation based on multi-time period, then prove the mechanism satisfies individual rationality and incentive compatibility. Finally, we try to use the real workload file to perform simulation experiments to verify the effectiveness of the improved online mechanism.

Keywords Virtual machine · Auction · Allocation · Mechanism design

1 Introduction

The cloud computing platform abstracts the physical computing resource into Virtual Machine instances, facilitating the management and pricing of resources. Users request for resources in the form of combination of different type of VMs to the cloud computing platform (auctioneer), then the auctioneer guarantee users right to use VM for a period of time and charge users to create cloud revenue.

Most of the auction mechanisms mentioned in the past work are off-line, users in this auction need to participate in multiple rounds to complete the task, and there is

L. Teng · H. Geng · Z. Yang · J. Zhu (✉)
School of Information Engineering, Yangzhou University,
Yangzhou 225127, China
e-mail: jwzhu@yzu.edu.cn

J. Zhu
Department of Computer Science and Technology,
University of Guelph, Guelph NIG2K8, Canada

© Springer International Publishing AG 2018
H. Lu and X. Xu (eds.), *Artificial Intelligence and Robotics*,
Studies in Computational Intelligence 752,
https://doi.org/10.1007/978-3-319-69877-9_7

53

no guarantee of coherent execution of the work, leading to the poor user experience. In auctions based on online provisioning and allocation of VMs, once the mechanism determines the winner set, it will ensure the continuous supply of the requested resources and the resource is automatically recovered when the task is completed. At the same time, online mechanism uses the critical point payment function to decide the payment and achieve economic win-win situation.

However, in most online auction, we generally ignore the difference in the number of competing users in various periods. In this paper, we roughly divide the time period into peak periods and trough periods. During the peak period, a large number of users initiate the virtual machine resource request to the auctioneer, resulting in resource constraints. The auctioneer need to set the time limit for the bidding in this time period and make efficient distribution. In addition, the problem of resource pricing for winners has been a hot topic among economists, in this paper, we hope to set the price reasonably for the auctioneer in order to bring more revenue.

2 Related Work

Scholars have developed a number of practical multi-item allocation algorithms: [1, 2] describes several practical tools for solving assignments, such as CPLEX in a single item combinational auction and a CABOB allocation algorithm in a multi-item portfolio auction. Huan Bai and Zili Zhang [3] improved CABOB allocation algorithm to choose the largest value set from a set of winners as the winner set. Sandholm [4] introduced the OR-of-XORs bidding language in the combined auction, and proposed a new tree search algorithm. Zurel [5] used a positive linear programming distribution function in the study of optimal allocation, which greatly maximizes the total value of the total revenue. However, these allocation methods are limited by the bid quantity involved in the auction, and difficult to find the price payment function correspondingly. Hoos [6] proposed a combination of auction under the local climbing algorithm, which is simple and practical, so that users can solve the large-scale bidding winner determination problem.

At present, most of pricing functions of the auction mechanism use the VCG mechanism [7, 8]. Dobzinski and Nisan [9, 10] studied the shortcomings of the VCG mechanism: when the allocation mechanism is the optimal allocation algorithm, the VCG payment function will no longer satisfies the motivational compatibility attributes, and improved the VCG mechanism, making the auction mechanism more individual rational. Mashayekhy [11, 12] studied the online real-time distribution of the auction mechanism, which simplifies steps that the cloud platform collecting the bidding, it can calculate the payment, allocate real-time resource and be applied in practical application environment. Nowadays, the VCG mechanism and the allocation algorithms have been widely used in other fields of AI [13–15].

3 Online Allocation Problem of Virtual Machine

3.1 An Auction Framework for VM Allocation

In the auction framework for virtual machine allocation, there are three components: cloud user, cloud platform, and cloud resource provider. Cloud users request for VM instance and offer bids to cloud platform. While cloud provider supply their resources in the form of various of VM instances, and sell them via allocation and pricing strategy,then use auction mechanism to auction unallocated resources after pricing. Cloud platform is a bridge to connect cloud users and the cloud provider. These modules work together to compete an auction. Then, we will describe the specific flow of this framework which is shown in Fig. 1.

On the one hand, cloud users offer bids to the platform in order to obtain the right of VM instances. The cloud platform collect bids and determine the winner set according to the allocation mechanism. In this paper, We have improved the mechanism by taking into account the changes in demand during peak and trough period so that the provider can make decisions reasonably, improve resource utilization, and bring higher profits. On the other hand, the cloud provider configure the resource and allocate instances to users, at the same time, the provider charge fees from users who have achieved resources according to the pricing function.

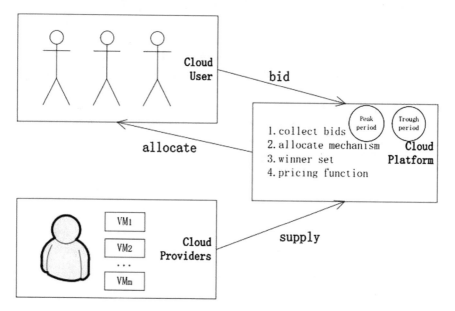

Fig. 1 The auction framework for virtual machine allocation

3.2   Concepts and Definitions

A cloud provider offers m different types of VM instances $VM_1, VM_2, \ldots, VM_m$ and each type of instance has D-dimensional resource, such as cores, memory, storage, etc. We assume that the cloud provider supply VM with a total amount of M, the number of VM for each type, denoted by $M_i(i = 1, 2 \ldots, m)$. VM_m can provide w_m^d number of d-type resources. For the d-type resources, the total amount resources a cloud provider can provide is C_d, the total number of virtual machine instance resources that the auction mechanism has to supply must be limited to C_d, that is $\sum_{i=1}^{m} M_i w_m^d \leq C_d, d \in D$.

Then, the winner determination problem is described as follows. There is a user set $U = \{U_1, \ldots, U_n\}$ of n users. User j, requests a bundle $S_j = (r_j^1, r_j^2, \ldots, r_j^m)$ of m types of VM instances, where r_j^i is the number of requested VM instances of type i and $\sigma_j^d = \sum_{i=1}^{m} r_j^m w_m^d$ is the total amount of each resource of type d that user j has requested. User j's request is denoted by $\theta_j = (S_j, a_j, l_j, d_j, v_j)$, where a_j is the arrival time of her request, l_j is the amount of time for which the requested bundle must be allocated, and d_j is the deadline for her job completion. We define $\delta_j = d_j - l_j$ as the deadline by which S_j must be allocated resource to user j in order for her job to complete its execution. User j values her requested bundle S_j by v_j, which is the maximum amount a user is willing to pay for obtaining resources if the resources can be allocated within time window $[a_j, \delta_j]$. If a user is not granted the allocation within this interval, her request will be declined and she will withdraw her request.

The objective of the auction mechanism is to determine the winner set. Given allocation vector $X = (x_1, x_2, \ldots, x_n)$, x_j represents the auction result of user j: $x_j = 1$ means that the resources is allocated to user j, and $x_j = 0$, otherwise. In order to better describe the assignment result of user j, we define the variable X_j^t, $X_j^t = 1$ means S_j is allocated to j at time t. The feasibility of the allocation to user j is indicated by y_j^t. $y_j^t = 1$ means the indicator parameter ensures that the allocation of the requested bundle is within time $a_j \leq t \leq \delta_j$. $P = (P_1, P_2, \ldots, P_n)$ is the payment rule that determines the amount each user must pay for the allocated bundles. We formulate the problem of online VM allocation and pricing as follows:

$$max \sum_{j=1}^{n} \sum_{t=1}^{T} v_j^t y_j^t X_j^t$$

4 Demand-Based Online Auction Mechanism for Virtual Machine Allocation

4.1 Design of Online Auction Mechanism

The online auction mechanism is different from the off-line mechanism in the step of collecting users' bids, it will accept user-initiated requests in time and decide the reasonable allocation. OVMCA takes an Event, the current allocation set $\mathcal{A}$, and the payment set $\mathcal{P}$ as input. The specific mechanism is given in Algorithm 1.

Algorithm 1 OVMCA Mechanism(Event,$\mathcal{A}$,$\mathcal{P}$)

1: $t \leftarrow$ Current time
2: $N^t \leftarrow \{\theta_j | j \in U$,the resouces j requested is not allocated$\}$
3: $\tilde{N}^t \leftarrow \{ \theta_j | j \in U$,(the resouces j requested has been allocated) $\wedge$ (her job is not finished) $\}$
4: **for all** $j \in U$ **do**
5:     **for all** $d \in D$ **do**
6:         $\sigma_j^d = \sum_{i=1}^{m} r_j^m w_m^d$
7:     **end for**
8: **end for**
9: **for all** $d \in D$ **do**
10:     $C_d^t \leftarrow \sum_{j | \theta_j \in \tilde{N}^t} \sigma_j^d$
11: **end for**
12: $C^t \leftarrow (C_1^t, \ldots, C_D^t)$
13: **if** $N^t = \emptyset$ or $C^t = 0$ **then**
14:     return
15: **end if**
16: $\mathcal{A}^t \leftarrow OVMAP - ALLOC(t, N^t, C^t)$
17: $\mathcal{A} \leftarrow \mathcal{A} \cup \mathcal{A}^t$
18: $\mathcal{L}^t \leftarrow \{\theta_j | \theta_j \in N^t \wedge (\theta_j, t) \notin \mathcal{A}^t\}$
19: $\mathcal{P} \leftarrow \mathcal{P} \cup \{v_j | (\theta_j, t) \in \mathcal{A}^t\}$
20: $\mathcal{P} \leftarrow OVMAP - PAY(t, N^t, L^t, A, P, C^t)$
21: Output $\mathcal{A}$,$\mathcal{P}$

In lines 1 to 11, OVMAP sets the current time to t and initializes four variables as follows: N^t, the set of requests of the users that have not been allocated: $N^t = \{\theta_j | a_j \leq t \wedge \neg(\exists t_j \leq t : (\theta_j, t_j) \in \mathcal{A})\}$; $\tilde{N}^t$, the set of requests of the users that have been allocated and their jobs have not finished yet. Formally, $\tilde{N}^t = \{\theta_j | \exists t_j < t : (\theta_j, t_j) \in \mathcal{A} \wedge t_j + l_j > t\}$; σ_j^d: the amount of d-type resource requested by user j; C_d^t: the available capacity of the resource d at time t.

The mechanism stores the resource capacities as a vector C^t at time t (line 12). Then, it proceeds only if resources and requests are available. The allocation function OVMAP-ALLOC returns $\mathcal{A}^t$, the set of users who would receive their requested bundles at time t (line 16–17). The mechanism then updates the overall allocation set $\mathcal{A}$. Then it counts the set $\mathcal{L}^t$ that bids are not available in this allocation, it helps the mechanism determine the payment of users. The payment of users in $\mathcal{A}^t$ are inserted into the payment set as their initial payment. The payment function OVMCP-PAY returns updated set $\mathcal{P}$ containing the payment of users at time t (line 19–20).

4.2 Properties of OVMCA Mechanism

Theorem 1. *OVMAC mechanism is incentive-compatible.*

Proof. We first show that the allocation algorithm OVMCA-ALLOC is monotonic. If user j wins by reporting v_j, then she will also win if she reports a more preferred request which $v'_j > v_j$. This is because the benchmark density h^t_j is unchanged and density value $f_j < f'_j$, the relative density g'_j is higher than $g_j = f_j - h^t_j$, which will win because of the high density. When counting the payment, we consider the bid density of user q from losing set where q is the index of user q appearing after user j based on the non-increasing order of the bid density. If user j reports a bid below the minimum value, she loses; otherwise she wins. This unique value is the critical payment for user j. Since the payment is the critical payment and the allocation function is monotonic, OVMAP is incentive-compatible.

Theorem 2. *OVMCA mechanism is individually rational.*

Proof. We consider user j as a winning user. We need to prove that if user j reports her true request, her utility is non-negative. This can be easily seen from the structure of the OVMACA mechanism. The payment for user j is set to $P_j = (h^t_j + g_{j'}) \cdot l_j \cdot \prod_{d \in D} \sigma^\alpha_j$, where user q is the user who would appears after user j in the decreasing order of the density metric. As a result, the utility of user j (i.e., *Utility* $= v_j - p_j$) is non-negative, that is she never incurs a loss, she would like to take apart in the auction.

5 Experimental Results and Analysis

We use real workload file to perform simulation experiments to verify the effectiveness of the proposed online mechanism. Our experimental data is taken from the workload log records of Grid Workloads Archive (GWA) and Parallel Workloads Archive (PWA). By comparing our improved algorithm with the conventional online distribution algorithm OVMPA, the advantages are proved.

5.1 Experimental Design

We selected six logs based on the availability of both recorded CPU and memory requests/usage, each log represents a series of request. We will mark these fields as

User j, the arrival time a_j, the use time to finish the job l_j, and the user's resource request S_j; we generate the user's valuation v_j by the random value between 1 and 10; the deadline is the random value between 3 and 6 multiplied by the job execution time l_j. In order to reflect the difference in the benchmark price of the time period, we set the basis price for the trough period to be 0 and the peak period be 0.5 respectively.

5.2 Analysis of Results

We use the variable-controlling approach to carry out two sets of simulation experiments to ensure that the consistency of all data except basis price. In Fig. 2, we compared the benefits of cloud provider between online mechanisms. We found that the improved mechanism obtain more revenue than typical online auctions. Through considerations of the density in peak and trough period, giving priority to the bids with smaller running time in the peak period will allow cloud resources to serve more cloud users, choosing the bid with long running time in the low period will reduce virtual machine consumption. Under the pricing scheme based on the price retention mechanism, the phenomenon that user use virtual machine resources without payment is reduced, which makes the auctioneer's income steadily improved.

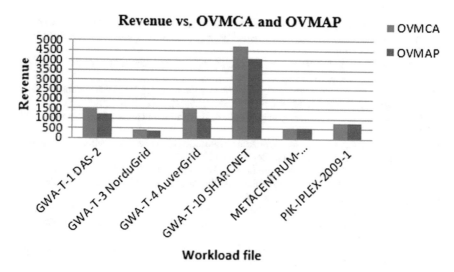

Fig. 2 Comparison result of social welfare

6 Conclusions

In this paper, we consider the change in the demand of peak and trough periods and try to improve the Online Virtual Machine Auction Mechanism, so that it can dynamically measure the bid density of users under different time and decide a more reasonable winner set. The experiment shows that this allocation mechanism can more effectively improve the resource utilization rate and bring higher profits for the auctioneer. In the future work, we can do research on the cloud resource basis price (ladder-type segment price function) for each time period, this continuous price function seems to be more powerful than others.

Acknowledgements Project supported by the National Nature Science Foundation of China (Grant No.61170201, No.61070133, No.61472344); Six-talent peaks project in Jiangsu Province (Grant No.2011-DZXX-032). Innovation Foundation for graduate students of Jiangsu Province (Grant No.CXLX12 0916), Jiangsu Science and Technology Project No. BY2015061-06BY2015061-08, Yangzhou Science and Technology Project No. SXT20140048, SXT20150014, SXT201510013, Natural Science Foundation of the Jiangsu Higher Education Institutions (Grant No.14KJB520041), Innovation Program for graduate students of Jiangsu Province (Grant No.SJZZ16_0261).

References

1. Bassamboo, A., Gupta, M., Juneja, S.: Efficient winner determination techniques for internet multi-unit auctions. In: Ifip Conference on Towards the E-Society: E-Commerce, E-Business, E-Government (2002)
2. Sandholm, T., et al.: CABOB: a fast optimal algorithm for winner determination in combinatorial auctions. Manag. Sci. **51**(3), 374–390 (2005)
3. Zheng, G., Lin, Z.C.: A winner determination algorithm for combinatorial auctions based on hybrid artificial fish swarm algorithm. Phys. Proc. **25**(22), 1666–1670 (2012)
4. Sandholm, T.: Algorithm for optimal winner determination in combinatorial auctions. Artif. Intell. **135**(1–2), 1–54 (2002)
5. Zurel, E., Nisan, N.: An efficient approximate allocation algorithm for combinatorial auctions. In: Acm Conference on Electronic Commerce (2001)
6. Hoos, H.H., Boutilier, C.: Solving combinatorial auctions using stochastic local search. In: Seventeenth National Conference on Artificial Intelligence and Twelfth Conference on Innovative Applications of Artificial Intelligence (2000)
7. Cavallo, R.: Optimal decision-making with minimal waste: strategyproof redistribution of VCG payments. In: International Joint Conference on Autonomous Agents and Multiagent Systems (2006)
8. Lahaie, S., Parkes, D.C.: On the communication requirements of verifying the VCG outcome. In: ACM Conference on Electronic Commerce (2008)
9. Dobzinski, S., Nisan, N.: Limitations of VCG-based mechanisms. In: ACM Symposium on Theory of Computing, San Diego, California, Usa, June (2007)
10. Nisan, N., Ronen, A.: Computationally feasible VCG mechanisms. In: ACM Conference on Electronic Commerce (2011)
11. Mashayekhy, L., et al.: Incentive-compatible online mechanisms for resource provisioning and allocation in clouds. In: IEEE International Conference on Cloud Computing (2014)
12. Mashayekhy, L., et al.: An online mechanism for resource allocation and pricing in clouds. IEEE Trans. Comput. **65**(4), 1172–1184 (2016)

13. Huimin, L., Li, Y., Zhang, Y., Chen, M., Serikawa, S., Kim, H.: Underwater optical image processing: a comprehensive review. Mob. Netw. Appl., 1–12 (2017)
14. Lu, H., Li, Y., Chen, M., Kim, H., Serikawa, S.: Brain intelligence: go beyond artificial intelligence. Mob. Netw. Appl. (2017)
15. Lu, H., Li, Y., Mu, S., Wang, D., Kim, H., Serikawa, S.: Motor anomaly detection for unmanned aerial vehicles using reinforcement learning. IEEE Internet of Things J., 1–8 (2017)

A Joint Hierarchy Model for Action Recognition Using Kinect

Qicheng Pei, Jianxin Chen, Lizheng Liu and Chenxuan Xi

Abstract In this paper, we proposed a joint hierarchy model to represent the motion of human according to the covariance feature of adjacent joints using Kinect. SVM is used for the action classification. Experimental results show that the proposed model improves the recognition accuracy with less computation complexity.

Keywords Human action recognition · SVM · Joint hierarchy
Skeleton

1 Introduction

Human action recognition has been a hot topic in computer vision due to its wide range of applications in human computer interactions (HCI). There are two types of recognition: image-based and skeleton-based. The image based recognition methods take videos as the input [1–3]. e.g. Blank et al. [4] took the video sequence as a space-time intensity volume, and defined some space-time features for action recognition.

The skeleton-based approaches use the skeleton information by motion capture equipment. Xia et al. [5] used the histograms of 3D joints information as dictionary

Q. Pei (✉) · J. Chen (✉) · L. Liu · C. Xi
Key Lab of Broadband Wireless Communication and Sensor Network Technology,
Nanjing University of Posts and Telecommunications, Ministry of Education,
Nanjing NY217025, China
e-mail: peiqicheng94@163.com

J. Chen
e-mail: chenjx@njupt.edu.cn

L. Liu
e-mail: liulizheng9312@163.com

C. Xi
e-mail: Sheldon_xi@outlook.com

© Springer International Publishing AG 2018
H. Lu and X. Xu (eds.), *Artificial Intelligence and Robotics*,
Studies in Computational Intelligence 752,
https://doi.org/10.1007/978-3-319-69877-9_8

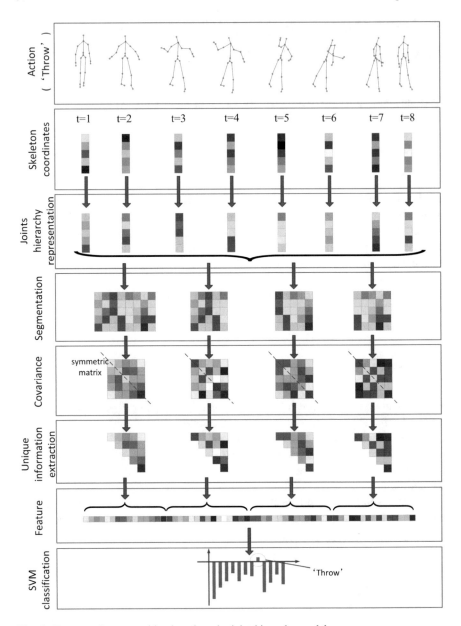

Fig. 1 Human action recognition based on the joint hierarchy model

for classification. Li et al. [6] recognized human actions from sequences of depth maps, and proposed a sampling scheme based on the projection of 3D points to reduce the recognition error. Yang et al. proposed an Eigenjoints skeletal representation [7], in which the pair-wise joint differences between frames are used for the classification. Yuan et al. [8] designed an approach based on a two-layer affinity propagation (AP) combined with Hidden Markov Model (HMM), which avoids the problem of random initialization of HMM. Lu et al. [9] extracted the action patterns in the framework via computing position offset of 3D skeletal body joints locally. Zhou et al. [10] proposed an extended label consistent K-SVD algorithm for learning the common and action-specific dictionaries, which is effective to extract key poses. Vemulapalli et al. [11] modeled human actions as curves in a Lie group. Li et al. [12] decreased the computing complexity by clustering skeleton joints. Zhu et al. [13] proposed a fusion scheme to combine the frame differences and pairwise distances of skeleton joint positions. Hussein et al. [14] concatenated the joint coordinates to extract temporal information with a segmentation strategy.

However, these skeletal representations suffer from amount of computation during the recognition procedure. In this paper, we propose a new recognition method based on the joint hierarchy model to combat this problem. Figure 1 depicts the framework of the proposed algorithm. When the subject starts to perform an action, Kinect would extract the skeleton information frame by frame. Then we perform the joint hierarchy model based skeletal representation. The depth of the color represents the numerical difference. The followed segmentation groups the colored squares in the same segmentation into blocks. Then the covariance is calculated on the blocks. In addition, the half part of the covariance matrix is taken out and formed in a row and SVM is used for classification. The following part is organized: The joint hierarchy model and skeleton representation are introduced in Sect. 2. In Sect. 3 experimental results verify the efficiency of the model. Section 4 concludes the paper.

2 A Joint Hierarchy Model

Kinect provides 20 joints of human body, and there are affiliations between these joints. For example, the left hand is connected with the left wrist. The Hand hooks up to the body via the wrist joint. Here the wrist joint holds a dominate position. According to this relation, we use parent/child to describe the adjacent joints. Let the dominate joint be the parent, and the other be the child. In human body, we define the spine joint as 'joint 1' according to the physical characteristic. Let 'joint 1' as the root and all joints are built into a binary tree as Fig. 2. In this tree, the first layer consists of the root joint, e.g. the 'joint 1'. The second layer consists of the shoulder and hip joints. The other layers are deduced according to this relation. Herein, all joints of human can be divided into 6 layers.

2.1 Action Sequence

Let the coordinate of joint as J, and the ith joint on layer l as $J_{i,l}$. We assume that the ith joint on layer $l-1$ as $J_{i,l-1}$. The bone vector between $J_{i,l}$ and $J_{i,l-1}$ as $D_{i,l}$. i belongs

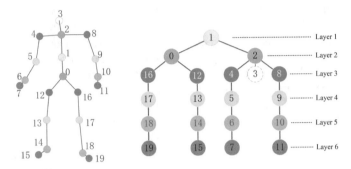

0 Hip center   1 Spine   2 Shoulder center   3 Head   4 Shoulder right   5 Elbow right
6 Wrist right   7 Hand right   8 Shoulder left   9 Elbow left   10 Wrist left   11 Hand left
12 Hip right   13 Knee right   14 Ankle right   15 Foot right   16 Hip left   17 Knee left
18 Ankle left   19 Foot left

Fig. 2 A joint hierarchy model

to {1,2} in layer 2 and {1, 2, 3, 4} in layer {3, 4, 5, 6}, l might be an integer between 2 and 6. Then we have

$$D_{i,l} = J_{i,l} - J_{i,l-1} \tag{1}$$

From Fig. 2 we find that there are 19 pairs of joints which meet such conditions. When we focus on the action of limbs, joint 3 might be ignored as it is the head joint. Then the number of interest pairs is 18. We connect all 18 vectors to form vector S in a frame t as Eq. (2). We have the representation of the complete action sequence as Eq. (3).

$$S^t = \left[D_{1,1}^{t\,\prime}, D_{2,1}^{t\,\prime}, \ldots, D_{i,l}^{t\,\prime}, \ldots, D_{4,6}^{t\,\prime} \right] \tag{2}$$

$$S = \left[S^1, S^2, \ldots, S^T \right] \tag{3}$$

The covariance would be calculated in the segmentation of S to generate the joint probability distribution. As described above, the number of D is 18, but the position information consists of the 3 axes coordinate, we form the coordinates of a joint as a row vector, so the length of $J_{i,l}^t$ and $D_{i,l}^t$ is 3, therefore, the length of S^t is 54. The superscript in Eq. (2) "′" means transposition. If the action includes T frames, the size of S is 54 × T.

2.2 Sequence Segmentation

Figure 3 depicts the segmentation for the frame sequence of complete action. The previous study found that the sequence is better to divide into 4 pieces [14]. Here

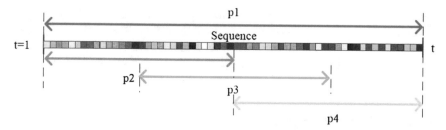

Fig. 3 The segmentation of joints' difference sequence

we adopt this segmentation. Firstly, the whole sequence is taken as one piece (noted as 'p1' in the figure, and so on for 'p2', 'p3', 'p4'). Moreover, it is segmented into several segments with the same length, e.g. p2, p3, and p4. Here p3 starts from the middle of p2 to the middle of p4. There is overlap between adjacent pieces.

For piece k, the covariance matrix is

$$C_k = E[(p_k - E(p_k))(p_k - E(p_k))'] \tag{4}$$

Here p_k denotes the kth piece, and E(.) is the expectation. The scale of the covariance matrix is N × N (Here N is 54), i.e.

$$C_k = \begin{bmatrix} c_{1,1} & c_{1,2} & \cdots & c_{1,N-1} & c_{1,N} \\ c_{2,1} & c_{2,2} & & c_{2,N-1} & c_{2,N} \\ \vdots & & \ddots & \vdots & \\ c_{N-1,1} & c_{N-1,2} & & c_{N-1,N-1} & c_{N-1,N} \\ c_{N,1} & c_{N,2} & \cdots & c_{N,N-1} & c_{N,N} \end{bmatrix} \tag{5}$$

For the kth part of the final eigenvector F (k belongs to {1, 2, 3, 4} as the sequence is divided into 4 pieces), we have

$$f_k = [c_{1,1}, c_{1,2}, \cdots, c_{1,N-1}, c_{1,N}, c_{2,2}, \cdots, \\ c_{2,N-1}, c_{2,N}, \cdots c_{N-1,N-1}, c_{N-1,N}, c_{N,N}] \tag{6}$$

According to the definition, the covariance matrix is symmetric. Then we concern on the above half triangle of the matrix. We form them into a vector f_k as in Eq. 6. The covariance matrix is 54 × 54, so the length of f_k is 1485. Then the eigenvector F is shown in Eq. (7).

$$F = [f_1, f_2, f_3, f_4] \tag{7}$$

The segmentation of sequence helps us extract the temporal information, because there may exist some actions which have same motion curves but reverse order of execution, such as the action 'push' and 'pull', while the covariance represents the distribution probability on the whole, so covariance cannot reveal the sequential order of the execution. While the segments formed in order like $f_1{\sim}f_3$ in F shows the order information of the process of the action. The covariance has been proved effective in skeleton-based action recognition [14].

As the length of f_k is 1485, the length of the final eigenvectors F is 5940.

3   Experiment Analysis

To see the efficiency of our proposed scheme, we let it work on two datasets: MSRC-12 [15] and MSR-Action3D [16]. In our experiments, the linear support vector machine (SVM) was used to train and test.

3.1   MSRC-12

MSRC-12 dataset contains 6244 gesture instance and records the coordinates of 20 skeleton joints. It has 30 subjects and consists of 12 actions. Four experiments are setup on this dataset. In test one, all of the samples were divided into three parts randomly, and one third for training and the rest for testing. While test two is the opposite of test one. The rest two are leave-one-out and 50% subjects-split. In leave-one-out cross validation we use the sequences of 29 subjects as training and the rest as testing. 50% Subjects-split: half of the actors are used for training and the rest for testing using ten different splits randomly. We use the gesture frames marked by the dataset as the starting and end point of actions. In other words, the sequences are divided by the marked frames: frames from number 0 to the first marked frame belong to the first action, and the frames from the next frame of the first marked frame to the second marked frame belong to the second actions and so on. The frames from the last marked frame to the last frame belongs to the last action in a sequence.

On MSRC-12, we found that the cross validation test (50% Subjects-split) is dramatically influenced by the 19th subject. It is 54.18 and 92.53% with and without the 19th subject. We checked the data of the 19th subject and found that it is not very stable, especially when the overlap of limbs happens. So in the following test we use the other 29 subjects as training while all subjects were used for testing.

Table 1 lists the results. In test one and test two, the proposed algorithm achieves the same results as that of Cov3DJ [14]. In the cross validation, the accuracy is closed to that of Cov3DJ, but 3% higher than that of the other algorithms. Although our scheme achieves the same performance as Cov3DJ, we avoid manually annotating the action sequence like Cov3DJ. We only use the marked frame provided by the dataset to annotate the starting and end. For the 50% subjects-split cross validation, our solution is best. In addition, our solution costs less computation as it uses less joints and shorter eigenvectors.

Table 1 MSRC-12 test result

MSRC-12	Test one	Test two	Leave-one-out-cross-subject-test	50% Subjects-split
Proposed	**98.0**	**98.7**	93.5	**92.5**
Cov3DJ [14]	97.9	98.7	**93.6**	91.7
Position offset + NBNN [9]	–	–	90.25	–
LC-KSVD [10]	–	–	90.22	–

3.2 MSR-Action3D

MSR-Action3D data set consists of 20 different actions performed by ten subjects. 557 action sequence would be used out of the 567 action sequences in all. This data set provides the coordinates of 20 joints like MSRC-12. This data set includes lots of similar actions which make it challenging. As in [6], we use the same experimental setup on MSR-A3D. The action classes were divided into three sets, each containing 8 action classes. Three experiment scenarios are setup on this dataset. In scenario one, all of samples were divided into three splits randomly, and one third for training and the rest for testing. Scenario two is the opposite of scenario one. In scenario three, the odd number of subjects were used for training and the even ones were used for testing, e.g. the sequences of subjects {1, 3, 5, 7, 9} are for training and {2, 4, 6, 8, 10} are for testing. Besides, we analyzed the other 3 different skeletal representations: Joints Positions (JP), Pairwise relative Joints Positions (RJP), and Angles between Joints (JA) [11].

Table 2 lists the results. In scenario one, our solution achieves 98.65% accuracy on AS3. It outperforms the other algorithms except that in [8]. Compare to that in [8], our scheme is much easier to build, and it achieves higher accuracy on AS1.

In test two, our solution works much well on all three sets. It achieved 98.63%, the highest on AS1. On AS2 our solution is little worse than that of Eigenjoints [7] about 2.7%. But on AS3, our solution achieved 100%, and we are the best on average.

In the cross subject test, our solution shows relative better stability, and it is better than that of [6–8], on AS1 and AS2. Even though the proposed is not the best in all the three subsets, our scheme achieves the highest accuracy in average.

Figure 4 shows the confusion matrix of cross-validation tests on all three sub-collections. Our solution achieved good results on most of the actions in AS1. For the actions of 'hand clap' and the 'horizontal arm wave', the accuracy is 100%. In AS2, the accuracies of 'two hand wave' and 'side boxing' also reach 100%. In AS3, our solution achieves best performance on 'forward kick', 'jogging', 'tennis swing' and 'golf swing'. Our solution is good for recognizing the actions like

Table 2 MSR-Action3D experiments results

	Test one				Test two				Cross validation			
	AS1	AS2	AS3	AVE	AS1	AS2	AS3	AVE	AS1	AS2	AS3	AVE
Proposed	89.73	86.84	98.65	91.74	**98.63**	96.05	**100**	**98.23**	86.67	79.46	90.09	**85.4**
Bag of 3D points [6]	89.5	89.0	96.3	91.6	93.4	92.9	96.3	94.2	72.9	71.9	**85.5**	74.7
HOJ3D [5]	**98.5**	**96.7**	93.5	**96.2**	98.6	97.9	94.9	97.2	**87.9**	85.5	63.5	79.0
Eigenjoints [7]	94.7	95.6	97.3	95.8	97.3	**98.7**	97.3	97.8	74.5	76.1	**96.4**	82.3
Two-layer AP + HMM [8]	89.0	88.2	**99.3**	92.17	97.2	92.0	100	96.4	83.8	75.0	93.7	84.17

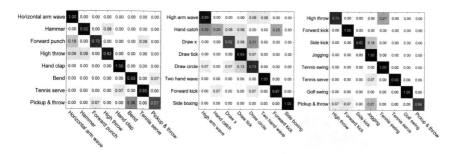

Fig. 4 Confusion matrix of cross-validation tests

Table 3 Cross validation results for several skeletal representation MSR-Action3d

	Ours	JA	JP	RJP
AS1	**86.67**	76.19	80.95	**86.67**
AS2	79.46	75.89	80.36	**81.25**
AS3	90.09	88.29	**90.99**	88.29
Average	**85.41**	80.12	84.10	85.40

Table 4 Cross validation time consumption for several skeletal representation MSR-Action3d

	Proposed	JA	JP	RJP
Data processing	**4.239689**	20.36988	4.902113	345.3124
Training	1.30495	2.865334	**1.211872**	262.6497
Prediction	**0.25954**	0.607116	0.290767	46.15931
Summation	**5.80418**	23.84233	6.404752	654.1214

'wave', 'swing' and 'kick', no matter the action is double-handed, single-handed or made by the leg. This is due to that the skeleton representation eliminates the redundant information between skeleton joints.

Table 3 lists the results compared with some skeletal representations referred in [11]. From it, we find that our solution works best on AS1. Although the RJP achieved the same accuracy as our scheme, it costs more computation. In AS2 and AS3, our scheme is best in average.

In addition, the proposed algorithm works with less computation complexity. Table 4 shows the computation (in seconds) used for the cross validation of MSRAction3D. The test environment is intel(R) Core(TM) i5-4590 CPU 3.3 Ghz with 16 GB RAM. The RJP skeletal representation achieved almost the same performance as our solution but the time used by RJP skeletal representation is 130 times.

4 Conclusion

In this paper, we proposed a joint hierarchy model for action recognition with Kinect, fusing the time series of skeletal joint motions and the relative dependence between joints. Firstly, We calculate the differences between joint and its parent according to the model definition. Then we employ a segmentation method for the frame, and calculate the covariance matrix in each segmentation. Finally we employ a SVM classifier to do the classification. In the future work, we would try to employ the clustering algorithm to improve the accuracy.

Acknowledgements This work was supported by the open research fund of Key Lab of Broadband Wireless Communication and Sensor Network Technology (Nanjing University of Posts and Telecommunications, Ministry of Education, NY217025), funding from Nanjing University of Posts and Telecommunications(NY217021), National Natural Science Foundation of China (Grant No. 61401228), China Postdoctoral Science Foundation (Grant No. 2015M581841), and Postdoctoral Science Foundation of Jiangsu Province (Grant No. 1501019A).

References

1. Lu, H., Li, Y.,. Chen, M., Kim, H., Serikawa, S.: Brain intelligence: go beyond artificial intelligence. Int. J. Comput. Sci. Comput Vis. Patt. Recogn. eprint arXiv:1706.01040 (2017). https://arxiv.org/abs/1706.01040
2. Lu, H., Li, B., Zhu, J., Li, Y., Xu, X., He, L., Li, X., Li, J., Serikawa, S.: Wound intensity correction and segmentation with convolutional neural networks. J. Concurrency Comput: Pract. Experience **29**(6) (2017). http://onlinelibrary.wiley.com/doi/10.1002/cpe.v29.6/issuetoc
3. Lu, H., Li, Y., Uemura, T., Ge, Z., Xu, X., He, L., Serikawa, S.: FDCNet: filtering deep convolutional network for marine organism classification. J. Multimedia Tools Appl. 1–14 (2017)
4. Blank, M., Gorelick, L., Shechtman, E. et al.: Actions as space-time shapes. Comput. Vis. ICCV 2005. In: Tenth IEEE International Conference on IEEE, vol. 2, pp. 1395–1402 (2005)
5. Xia, L., Chen, C.C., Aggarwal, J.K.: View invariant human action recognition using histograms of 3D joints. In: IEEE Computer Society Conference on Computer Vision and Pattern Recognition Workshops, pp. 20–27. IEEE (2012)
6. Li, W., Zhang, Z., Liu, Z.: Action recognition based on a bag of 3D points. In: IEEE Computer Society Conference on Computer Vision and Pattern Recognition-Workshops, pp. 9–14. IEEE (2010)
7. Yang, X., Tian, Y.L.: EigenJoints-based action recognition using naïve-bayes-nearest-neighbor **38**(3c), 14–19 (2012)
8. Yuan, M., Chen, E., Gao, L.: Posture selection based on two-layer AP with application to human action recognition using HMM. In: IEEE International Symposium on Multimedia, pp. 359–364. IEEE Computer Society (2016)
9. Lu, G., Zhou, Y., Li, X., et al.: Efficient action recognition via local position offset of 3D skeletal body joints. Multimedia Tools Appl **75**(6), 3479–3494 (2016)
10. Zhou, L., Li, W., Zhang, Y., et al.: Discriminative key pose extraction using extended LC-KSVD for action recognition. In: International Conference on Digital Image Computing: Techniques and Applications, pp. 1–8. IEEE (2014)

11. Vemulapalli, R., Arrate. F., Chellappa, R.: Human action recognition by representing 3D skeletons as points in a Lie group. In: IEEE Conference on Computer Vision and Pattern Recognition, pp. 588–595. IEEE Computer Society (2014)
12. Li, J., Chen, J., Sun, L.: Joint motion similarity (JMS)-based human action recognition using kinect. In: International Conference Digital Image Computing: Techniques and Applications (DICTA), pp. 1–8. IEEE (2016)
13. Zhu, Y., Chen, W., Guo, G.: Fusing spatiotemporal features and joints for 3D action recognition. In: Computer Vision and Pattern Recognition Workshops, pp. 486–491. IEEE (2013)
14. Hussein, M.E., Torki, M., Gowayyed, M.A., et al.: Human action recognition using a temporal hierarchy of covariance descriptors on 3D joint locations. In: Proceedings of the Twenty-Third International Joint Conference on Artificial Intelligence, pp. 2462–2472 (2013)
15. Fothergill, S., Mentis, H., Kohli, P., et al.: Instructing people for training gestural interactive systems. In: Sigchi Conference on Human Factors in Computing Systems, pp. 1737–1746. ACM (2012)
16. Li, W., Zhang, Z., Liu, Z.: Action recognition based on a bag of 3D points. In: Computer Vision and Pattern Recognition Workshops, pp. 9–14. IEEE (2010)
17. Shotton, J., Fitzgibbon, A.W., Cook, M., Sharp, T., Finocchio, M., Moore, R., Kipman, A., Blake, A.: Real-time human pose recognition in parts from single depth images. In: CVPR, pp. 1297–1304 (2011)

QoS-Based Medical Program Evolution

Yongzhong Cao, Junwu Zhu, Chen Shi and Yalu Guo

Abstract Medical path varies due to the changes in external factors, and the key to the changes is the decision-making in symptomatic treatment by medical experts; how to make full use of historical medical data and recommend high quality treatment options to new cases under the current circumstance of data explosion, is the research focus of this paper. This paper first presents the standard for medical path based on cloud platform, going into the definition of model of medical services; finally, the medical optimization factor is given to realize the evolution of online medical program of a disease based on QoS.

Keywords Medical path · QoS · Medical cloud

1 Introduction

Medical program is the treatment models and treatment procedures established for a disease following the empirical data of contemporary medicine and the clinical experience of doctors. Based on consideration for overall situation of health care, it is individualized, data dispersed and relatively regular. Large amounts of medical data are stored in the subsystems of each hospital, and the data format is not uniform. In the cloud computing framework, how to make full use of these data is the hot spot concerned by people Thompson [1].

In recent years, service composition has been gradually applied to data management for medical program. Service composition is the technology by reusing the atomic services that exist in the system as components, which is to select specific services from the existing services and combine them into a new service flow, so that to meet the needs of users and adapt to changes in the environment. Another

Y. Cao (✉) · J. Zhu · C. Shi · Y. Guo
Department of Computer Science and Technology,
Yangzhou University, Jiangsu 225000, China
e-mail: caoyz@yzu.edu.cn

© Springer International Publishing AG 2018
H. Lu and X. Xu (eds.), *Artificial Intelligence and Robotics*,
Studies in Computational Intelligence 752,
https://doi.org/10.1007/978-3-319-69877-9_9

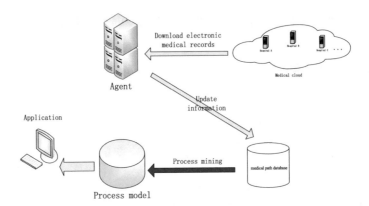

Fig. 1 Medical service flow evolution based on process mining

approach is to extend the definition of original workflow to adapt it to the dynamic, distributed and loosely coupled services.

In Fig. 1, through the cloud computing model,this paper defines the medical service flow based on definition of the standard for medical path, devoting to construct the initiative evolution of the medical program for disease in medical cloud, so as to minimize and avoid the arbitrariness of the treatment process and improve the quality of medical services.

2 Standards for Medical Path

Currently, a patient's medical path is dispersed in various subsystems of the hospital, such as: the patient's electronic medical record, the patient's hospitalization record, and so on. Based on the cloud platform system, in order to achieve data sharing, a unified path standard must be defined in its structure. The system needs to be manually or semi-manually organized into a medical path that can be shared among machines [2].

Definition 1 Medical path: real-time recording of the global information for the whole medical process of a disease, as well as the running of sub-service instances. Global information includes patient's name, date of hospitalization and postoperative evaluation; sub-service information includes serial number of the service instance (case_id), name of service (service_name), work order of the service in the instance (case_order), time of beginning (begin_time), time of ending (end_time) and cost of the service (service_cost). As the communications between various medical sub-services, as well as between service and flow Management Agent [3] are all conducted through XES documents, the format for XES document of Serviceflow_log is defined as follows:

```
<?xml version = "1.0" encoding = "UTF-8" ?>
<log  xes.version = "1.0"  xes.features = "nested-attributes"  openxes.version =
"1.0RC7">
<extension name = "patient _name " keys = "concept: patientname"/>
<extension name = " evaluation " prefix = "concept: evaluation "/>
<classifier name = " service_name " keys = "concept:name"/>
<classifier name = " case_order " keys = "concept:name lifecycle:transition"/>
<classifier name = " begin_time " keys = "concept:name"/>
<classifier name = " end_time " keys = "concept:name lifecycle:transition"/>
<classifier name = " service_cost " keys = "concept:name"/>
```

This form of service flow logs is stored on the relevant server of cloud platform, and these logs are used by the management node for real-time analysis of medical path.

Definition 2 Dependencies in Flow Log

Let w be a medical path log, and T be the corresponding task set, for a, b∈T, there are four kinds of relations: direct dependency, pure dependency, no dependency and interdependency [4].

The various dependencies of the sub-services in the medical service flow can be obtained through flow service log mining.

The structure of the service flow may change during the running period, whether the changed service flow is consistent with the original service flow, finding these dependencies from the medical log above is the basis for recreating the medical path during running period, and is also the basis for the evolution of medical path mentioned later.

3 Medical Service Flow Model

A Petri net system consists of places and transitions, and the dependencies among them are expressed on directed arcs. This formalized and graphical approach makes it more suitable for describing asynchronous concurrent systems. When the system correctly describes its transition, Petri net theory can guarantee that the whole system is achievable, and the formalized and graphical approach of Petri net facilitates the modeling of medical service flows.

By using Petri net, Alast has defined the workflow net [5] as follows:

Definition 3 Workflow Net (WF_net) is a Petri net PN (P, T, f), if and only if:

(1) PN has one source place i∈p, so that i = Φ.
(2) PN has one sink place o∈p, so that o = Φ, o does not have to be unique, because a workflow can have multiple exit points.
(3) Every node x∈P∪T is on a path from i to o.

The transitions in T represent the subtasks in the workflow, and the dependencies among subtasks (f) are represented by the connection between places. The relation between subtasks among the flow can be sequential, parallel or selective.

The workflow described in the above definition is static and cannot adapt to the dynamic and loosely coupled medical path. The function, which the dynamic adaptive workflow should have, has been summarized by Kammer in documentation. The main points are as follows:

Flow structures can be defined dynamically;
Flow activities, resources can be selected dynamically;
Flow structure evolution can be done according to the running instances of flow.

But the formalized definition of dynamic service flow has not been given, in order to achieve the above functions, the author has carried out expansion to dynamic feasible replacement mapping set, and the DWSF_net after the expansion is as follows:

Definition 4 DWSF_net is a five-tuple (P, T, f, Tr, r) if and only if:

(1) P, T and f in DWSF_net form a WF_net (P, T, f).
(2) Tr is a feasible alternative set of sub-services, for $\forall tr \in Tr$, $\exists t \in T$, so that tr = t; that is, t can be replaced by tr functionally.
(3) r is a mapping from Tr to t.$\forall t \in T$, the alternative set of t can be expressed as $\{x | x = r(t) \wedge x \in Tr\}$.

This allows that, after the completion of feasible replacement or supplement to the DWSF_net, (P, T, f) in DWSF_net is still a WF_net.

The DWSF_net corresponding to Fig. 1 can be expressed as follows:

In Fig. 2, the expression of DWSF_net adds the alternative service denoted by a hashed line and represent the mapping r between the service and its feasible alternative services; for the sake of simplicity of the graph, only alternative services for A and H are given here, for example, A1 and A2 are alternative services for A.

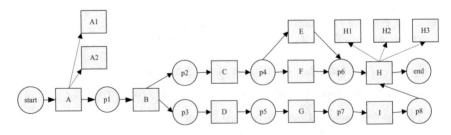

Fig. 2 DWSF_net example

4   QoS Model for Medical Service Flow

Postoperative evaluation of medical services for a disease can be derived from these areas, such as the patient's medical expenses, days of hospitalization, clinical examination results, patient satisfaction rate, the incidence of complications and re-hospitalization rate. The following aspects can be used to examine the service quality of medical service flow [6].

Service Cost: the sum of service costs for sub-service flows (Ω). It shall be noted that the resource costs of sub-services may vary at different time periods. Cost can be defined as:

$$Cost(\Omega) = \sum_{i=1}^{n} (\sum_{k=1}^{l} (\sum_{j=1}^{m} UT_j(t_{ik}) \times P(t_{ikj}))/l) \tag{1}$$

where i is the sub-service number, k is the number of cases executed by sub-services, $UT_j(t_i)$ is the amount of resource used or not used by sub-service t_{ij} in a single case:

$$UT_j(t_i) = \begin{cases} 1 & used \\ 0 & nonused \end{cases} \tag{2}$$

P (t_{ij}) is the unit price by using resource j.

Service Time: the time taken by sub-service flow (Ω) to complete a service. Due to external or internal reasons, the time for completion of each sub-service is not the same. For example: the patient's own circumstances, care and other factors may all affect the time for actual implementation of the task. Time can be defined as:

$$Time(\Omega) = \sum_{i=1}^{n} (ET_i - ST_i)/n \tag{3}$$

where ET is the moment on which the service is completed, ST is the startup time of the service. Formula 3 reflect the average time cost for the completion of this task instance by the sub-service flow. It also basically reflects the time quality of the service.

Service Rating SC: the sum of weighted credit for sub-service flows (Ω). In the provision of services by sub-service, their success rate will be recorded by the Management Agent of the system, such recording should have a process to reflect the credit of sub-services from scratch, from one-sided to comprehensive. SC can be defined as:

$$SC = \sum_i \lambda_i SA_i \tag{4}$$

where SA_i is the patient evaluation value of the medical path, n is the sum of the number of tasks undertaken since the credit file was established, λ_i is the weight of the evaluation for the medical path service.

The treatment effect refers to executing the treatment on the patients, which is compared with before treatment the patient's level of health. To quantify the extent of recovery during treatment, two detection values are first given:

Initial detection value $T_f = [t_{1_f}, t_{2_f}, t_{3_f}, \ldots, t_{n_f}]$ stands for the first admission test results, such as blood routine, biochemical liver function etc.

Final test value $T_l = [t_{1_l}, t_{2_l}, t_{3_l}, \ldots, t_{n_l}]$ represents the test results at discharge.

$$\text{Health indicators range for each test result:} \quad H = \begin{bmatrix} h_{1_{min}}, h_{1_{max}} \\ h_{2_{min}}, h_{2_{max}} \\ h_{3_{min}}, h_{3_{max}} \\ \ldots \\ h_{n_{min}}, h_{n_{max}} \end{bmatrix} \quad (5)$$

Among them, $h_{i_{min}}$ represents the lowest level of health, and $h_{i_{max}}$ represents the highest level of health.

Health distance is defined as follows:

$$DisH = \sqrt{\sum_{i=1}^{n} \left(\frac{t_i - \frac{h_{i_{max}} - h_{i_{min}}}{2}}{s_i} \right)^2} \quad (6)$$

The above four measures are developed in combination with the characteristics of medical services. It takes into account local sub-flow evaluation and global comprehensive evaluation.

5   QoS-Based Service Flow Evolution

Service cost, service time, service evaluation and health distance constitute the service flow measures of this paper, however, the types and values of these four measures are not the same, and must be standardized to carry out unified calculation.

Service cost, service time, health distance and service quality are inversely proportional, that is, the higher the service cost, the longer the service time, the longer the health distance, the worse the corresponding service quality; service evaluation and quality are directly proportional to each other, the higher the service evaluation, the better the corresponding service quality. In this way, the QoS calculation of the authorization occurrence sequence can be obtained by analyzing the service flow operation log.

After the QoS analysis of the sub-service flow is completed, a QoS matrix can be established for the entire service flow to facilitate the real-time evaluation of the entire service flow.

$$
\begin{bmatrix}
Q_{00}Q_{01} \cdots Q_{0i} \cdots Q_{on} \\
Q_{10}Q_{11} \cdots Q_{1i} \cdots Q_{1n} \\
\cdots \\
Q_{j0}Q_{j1} \cdots Q_{ji} \cdots Q_{jn} \\
\cdots \\
Q_{m0}Q_{m1} \cdots Q_{mi} \cdots Q_{mn}
\end{bmatrix} = QoS(Dwsf\_net) \qquad (7)
$$

With the updating of medical knowledge and the use of new drugs, medical path is dynamically changing; in order to ensure that the system can adapt to the changing external environment, the system must be aware of external changes and automatically adjust its structure to ensure its advanced nature. This kind of automatic structure optimization and adjustment to system is called service flow evolution. Evolution can be divided into several methods, such as replacement, split and merge [7].

In the service flow, the evolution by using a candidate service to replace a service is called replacement and is denoted by $\leftrightarrow$. For example, $a \leftrightarrow b$ indicates that service a is replaced by service b.

Set there is sub-service A_1, and its feasible alternative set is R_{A1}, lower limit of the rating threshold corresponding the sub-service is Min, upper limit is Max.

The rules for replacement are as follows:

$$
S_{SF}(A_1) < Min \wedge \exists t \in R_{A1} \wedge S_{SF}(t) > Min => t \leftrightarrow A_1 \qquad (Rule\ 1)
$$

$$
S_{SF}(A_1) > Min \wedge \exists t \in R_{A1} \wedge S_{SF}(t) > S_{SF}(A_1) => t \leftrightarrow A_1 \qquad (Rule\ 2)
$$

$$
S_{SF}(A_1) > Max \wedge \exists t \in R_{A1} \wedge S_{SF}(t) < Max => t \leftrightarrow A_1 \qquad (Rule\ 3)
$$

$$
\overrightarrow{gradS_{SF}}(A_1) = \overset{n}{\underset{i=1}{Max}}\,\overrightarrow{gradS_{SF}}(A_i) \wedge \exists t \in R_{A_1} \wedge S_{SF}(t)
$$

$$
< S_{SF}(A_1) \Rightarrow t \leftrightarrow A_1 \qquad (Rule\ 4)
$$

$$
\overrightarrow{gradS_{SF}}(A_1) = \overset{n}{\underset{i=1}{Min}}\,\overrightarrow{gradS_{SF}}(A_i) \wedge \exists t \in R_{A_1} \wedge S_{SF}(t)
$$

$$
> S_{SF}(A_1) \Rightarrow t \leftrightarrow A_1 \qquad (Rule\ 5)
$$

Figure 3 shows that the DWSF_net changes after the replacement of local sub-service flow.

The evolution, for which, sub-service or sub-service sequence is replaced by one or more alternative services to undertake together with the original service, is called split. The new sequences and the original service maintain parallel.

Set there is sub-service A_1, and its feasible alternative set is R_{A1}, lower limit of the rating threshold corresponding the sub-service is Min.

The rules for Segmentation are as follows:

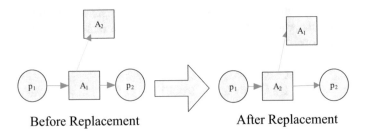

Fig. 3 Replacement of local DWSF_net

$$S_{SF}(A_1) < Min \wedge \exists t \in R_{A1} \wedge S_{SF}(A_1 + t) > Min => t + A_1 \leftrightarrow A_1 \quad \text{(Rule 6)}$$

$$S_{SF}(A_1) > Min \wedge \exists t \in R_{A1} \wedge S_{SF}(t + A_1) > S_{SF}(A_1) => t + A_1 \leftrightarrow A_1 \quad \text{(Rule 7)}$$

$$\overrightarrow{gradS_{SF}}(A_1) = \underset{i=1}{\overset{n}{Min}} \overrightarrow{gradS_{SF}}(A_i) \wedge \exists t \in R_{A_1} \wedge S_{SF}(t + A_1)$$

$$> S_{SF}(A_1) \Rightarrow t + A_1 \leftrightarrow A_1 \quad \text{(Rule 8)}$$

Figure 4 shows the DWSF_net changes after the split of local sub-service flow.

The evolution by removal of alternative services from quality-degraded region to improve comprehensive rating is called merge. That is, the task which was previously undertaken by a number of services will be completed by part of the services.

Set there is sub-service region undertaken by A_1 and t, and its feasible alternative set is R_{A1}, upper limit and lower limit of the rating threshold corresponding to the sub-service are Max and Min respectively.

The rules for Incorporation are as follows:

$$S_{SF}(A1 + t) > Max \wedge S_{SF}(A1) < Max \wedge S_{SF}(A1) > Min => A_1 \leftrightarrow t + A_1 \quad \text{(Rule 9)}$$

$$S_{SF}(A_1 + t) > Min \wedge \exists t \in R_{A1} \wedge S_{SF}(A_1) > S_{SF}(t + A_1) => A_1 \leftrightarrow t + A_1 \quad \text{(Rule 10)}$$

$$\overrightarrow{gradS_{SF}}(A_1 + t) = \underset{i=1}{\overset{n}{Max}} \overrightarrow{gradS_{SF}}(A_i + t) \wedge \exists t \in R_{A1} \wedge S_{SF}(A_1)$$

$$> S_{SF}(A_1 + t) \Rightarrow A_1 \leftrightarrow A_1 + t \quad \text{(Rule 11)}$$

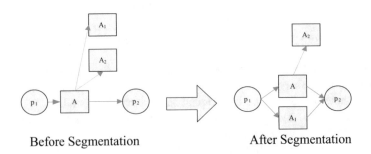

Fig. 4 Segmentation of local DWSF_net

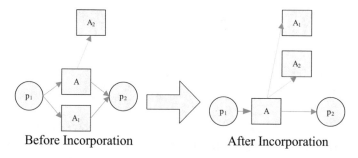

Before Incorporation   After Incorporation

Fig. 5 Incorporation of local DWSF_net

Figure 5 shows the DWSF_net changes after the merge of local sub-service flow.

Then how to select the appropriate candidate services in a large number of candidate sets to complete the evolution becomes the issue that must be resolved. In this paper, an improved genetic algorithm is used to select the best candidates in the candidate set to find out the optimal solution to undertake the current service [8, 9].

In the selection of the initial population, the optimal combination of the last calculation is added to the initial population, so as to improve the convergence speed. This is based on the user's willingness and task node to remain relatively stable over a period of time. It is in line with the opening and distribution characteristics of medical services under cloud computing environment [10–13].

With genetic algorithm, you can select in real time the recommended medical path that is in line with the user and system requirements, while taking into account the evaluation to the treatment by patients.

We used a medical path example of 56 patients in a hospital for cardia cancer as the experimental object, with the support of the definition of unified medical ontology, we collated the standard document of definition 1 by semi-manual method, adopting binary coding rule and PEV algorithm, randomly generating several initial populations of different service classes with very few iterations, and

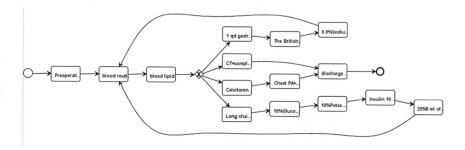

Fig. 6 Medical program mining results

the algorithm can find the optimal path very soon. By genetic algorithm based on process mining, we can get a medical path such as Fig. 6.

6 Conclusion

In order to solve the issue for service quality and service cost of medical path, this paper introduces QoS analysis mechanism and selects the medical path service according to user's willingness and environmental requirement in real time through the evolution of service flow; it can support management Agent to realize the online medical path automatic access, and has strong adaptability for the cloud computing environment with scattered nodes and dynamic changes in optional services.

References

1. Thompson, S.: From value assessment to value cocreation: informing clinical decision-making with medical claims data. Big Data **4**, 141–147 (2015)
2. van der Aalst, W.M.P.: Extracting event data from databases to unleash process mining. In: Management for Professionals, vol.2, pp. 105–128
3. Valentijn, P.P., Biermann, C., Bruijnzeels, M.A.: Value-based integrated (renal) care: setting a development agenda for research and implementation strategies **16**, 330 (2016)
4. Cao, Y., Ding, Q., Li, B.: Dynamic integration of service flows based on fairness. Comput. Integr. Manuf. Syst. **14**(12), 2457–2462 (2008)
5. van der Aalst, W.M.P.: The application of Petri Nets to workflow management. J. Circuits Syst. Comput. **8**(1), 21–66 (1998)
6. Shah, S.M., Zaidi, S., Ahmed, J., Rehman, S.U.: Motivation and retention of physicians in primary healthcare facilities: a qualitative study from Abbottabad, Pakistan. Int. J. Health Policy Manag.-IJHPM. **5**(8), 467–475 (2016)
7. Altiparmak, F., Gen, M., Lin, L., et al.: A genetic algorithm approach for multi-objective optimization of supply chain networks. Comput. Ind. Eng. **51**(1), 196–215 (2006)
8. Thriskos, P., Zintzaras, E., Germenis, A.: DHLAS: a web-based information system for statistical genetic analysis of HLA population data. Comput. Methods Programs Biomed. **85**(3), 267–272 (2007)
9. Agrawal, R., Imieliski, T.: A swai mining association rules between sets of items in large database. In: Proceedings of ACM SIGMOD International Conference on Management of Data (SIGMOD'93), pp. 207–216
10. Lu, H., Li, Y., Chen, M., Kim, H., Serikawa, S.: Brain intelligence: go beyond artificial intelligence, Mob. Networks Appl. 1–10 (2017)
11. Lu, H., Li, Y., Mu, S., Wang, D., Kim, H., Serikawa, S.: Motor anomaly detection for unmanned aerial vehicles using reinforcement learning. IEEE Internet Things J. https://doi.org/10.1109/JIOT.2017.2737479
12. Lu, H., Li, B., Zhu, J., Li, Y., He, L., Li, J., Serikawa, S.: Wound intensity correction and segmentation with convolutional neural networks. Concurrency Comput.: Pract. Experience. https://doi.org/10.1002/cpe.3927
13. Serikawa, S., Lu, H.: Underwater image dehazing using joint trilateral filter. Comput. Electr. Eng. **40**(1), 41–50 (2014)

An Improved 3D Surface Reconstruction Method Based on Three Wavelength Phase Shift Profilometry

Mingjun Ding, Jiangtao Xi, Guangxu Li, Limei Song
and Philip O. Ogunbona

Abstract In order to reduce the noise points when measure the shape of 3D object using Phase Shift Profilometry (PSP) methods, in this paper we propose a novel Three Wave Length PSP (TWPSP) method. Firstly, the direct problems of equivalent wavelength and unwrapped phase are analyzed. Then, the solution of unwrapping method for TWPSP is derived. Finally, based on the global phase filtering the phase noises are reduced. We simulated and compared the proposed TWPSP and the classical TWPSP. Experiential results shown that the noises are greatly restrained. Since the proposed method does not need to calculate the equivalent phase maps, the implement is improved than the classical TWPSP.

Keywords Phase shift profilometry · Denoising · Multi-wavelengths
Phase unwrapping

M. Ding · J. Xi · G. Li (✉) · P.O. Ogunbona
School of Electronic and Information Engineering,
Tianjin Polytechnic University, Tianjin, China
e-mail: liguangxu@tjpu.edu.cn

M. Ding
e-mail: dingmingjun@tjpu.edu.cn

J. Xi
e-mail: jiangtao@uow.edu.au

P.O. Ogunbona
e-mail: philip_ogunbona@uow.edu.au

J. Xi · P.O. Ogunbona
School of Electrical, Computer and Telecommunications Engineering,
University of Wollongong, Wollongong, NSW, Australia

L. Song
School of Electrical Engineering and Automation,
Tianjin Polytechnic University, Tianjin, China
e-mail: songlimei@tjpu.edu.cn

© Springer International Publishing AG 2018
H. Lu and X. Xu (eds.), *Artificial Intelligence and Robotics*,
Studies in Computational Intelligence 752,
https://doi.org/10.1007/978-3-319-69877-9_10

85

1 Introduction

1.1 *Fringe Pattern Profilometry Technology*

Fringe pattern profilometry (FPP) is an effective technology for non-contact measurement of 3D shapes. Phase shift profilometry (PSP) is widely used to reconstruct the 3D shape approach by advantages of its implement speed, reconstruct resolution and it is less sensitive to the surface reflectivity variations [1]. Many different algorithms have been developed, e.g. spatial phase unwrapping [2] and temporal phase unwrapping methods [3]. The temporal methods are more robust, but they are less efficient in terms of speed due to the use of multiple pattern images.

The temporal phase unwrapping techniques include the Coding PSP method and the Multi Wavelength PSP (MWPSP) method [4]. Chen et al. [5] described a technique to deal with the 2π phase ambiguity problem through utilizing two properly chosen frequency gratings to synthesize an equivalent wavelength grating. Towers et al. [6, 7] proposed an optimal frequency selection for the multiple wavelength method. They then verified that an optimal three wavelength for 3D shape measurement. Li et al. [8] presented a phase unwrapping approach based on multiple wavelength scanning for digital holographic microscopy, which unwrapped the ambiguous phase image layer by layer by synthesizing the extracted continuous components from a set of multiple phase images obtained by varying the optical wavelength. Ding et al. [9] proposed an approach to recover absolute phase maps of two fringe patterns with selected frequencies, they also provided a guide line for frequency selection. Long et al. [10] proposed a method which can be applied for any two fringe patterns with different fringe wavelengths by selecting the fringe waveform lengths but using the two frequencies to describe the two fringe patterns.

Due to the existence of noise in the environment when we project the fringe patterns, multiple intermediate fringe patterns must be used, and error points happened at the results of the unwrapping. Consequently, we developed a phase unwrapping algorithm based on three wavelength phase shift profilometry (TWPSP) method [11]. However, it still needs to calculate the equivalent phases through the equivalent wavelengths. The noise associated with the wrapped phase map will propagate to the unwrapped ones, leading to errors in the recovered phase maps. This technique could also be utilized for 3D underwater scene reconstruction [12–16].

In this paper, we present a new phase unwrapping approach based on the use of three fringe patterns with different wavelengths to suppress noises further. Compared with the existing method in [11], the proposed method does not need to calculate the equivalent phases. As the noise propagation and enhancement are avoided, the resulting unwrapped phase is less influenced by the noise, therefore leading to improved phase unwrapping performance.

1.2 Three Wave Length PSP Principal

λ_1, λ_2 and λ_3 denotes as the wavelengths of laser separately. $\phi_1(x, y)$, $\phi_2(x, y)$ and $\phi_3(x, y)$ are presented the corresponding wrapped phase maps. Phase maps $\phi_{12}(x, y)$, $\phi_{23}(x, y)$ and $\phi_{123}(x, y)$ can be obtained by the equivalent wavelengths λ_{12}, λ_{23} and λ_{123},

$$\lambda_{12} = \left|\frac{\lambda_1 \times \lambda_2}{\lambda_1 - \lambda_2}\right|, \ \lambda_{23} = \left|\frac{\lambda_2 \times \lambda_3}{\lambda_2 - \lambda_3}\right|, \ \lambda_{123} = \left|\frac{\lambda_{12} \times \lambda_{23}}{\lambda_{12} - \lambda_{23}}\right|. \tag{1}$$

According to the four steps PSP algorithm, four pattern images are projected to the object. And we can get the captured pattern as:

$$I_k(x, y) = a(x, y) + b(x, y) \sin\left(\phi(x, y) + \frac{\pi \times k}{2}\right), \ k = 0, 1, 2, 3. \tag{2}$$

where $a(x, y)$ notes as the background intensity; $b(x, y)$ is the intensity modulation, $\phi(x, y)$ is the wrapped phase. We have proved that the wrapped phase value $\phi i(x, y)$ can be calculated as

$$\phi_i(x, y) = \arctan\left(\frac{I_0(x, y) - I_2(x, y)}{I_1(x, y) - I_3(x, y)}\right), \ i = 1, 2, 3. \tag{3}$$

From Eqs. (1) and (3), the phase maps $\phi_{12}(x, y)$, $\phi_{23}(x, y)$ and $\phi_{123}(x, y)$ become to

$$\phi_{12}(x, y) = \arctan\left(\frac{\sin(\phi_2(x, y) - \phi_1(x, y))}{\cos(\phi_2(x, y) - \phi_1(x, y))}\right)$$

$$\phi_{23}(x, y) = \arctan\left(\frac{\sin(\phi_3(x, y) - \phi_2(x, y))}{\cos(\phi_3(x, y) - \phi_2(x, y))}\right) \tag{4}$$

$$\phi_{123}(x, y) = \arctan\left(\frac{\sin(\phi_{12}(x, y) - \phi_{23}(x, y))}{\cos(\phi_{12}(x, y) - \phi_{23}(x, y))}\right).$$

Support that the weight of fringe strips is W pixels and the direction is vertical. If the equivalent wavelength $\lambda_{123} \geq W$, $\phi_{123}(x, y)$ varies continuously from $[0, 2\pi)$, the figures of the wrapped phase can be expressed definitely. In this case, phase unwrapping is not required, since the absolute phase map is as same as the wrapped one. Although $\phi_{123}(x, y)$ does not exhibit 2π discontinuities, the reconstruction of the 3D surface from $\phi_{123}(x, y)$ directly will not achieve an ideal result as the long wavelength usually causes measurement accuracy becoming poor. Generally, if $\phi_1(x, y)$ can be unwrapped according to the equivalent phases, the result will be better. However, on $\phi_1(x, y)$, there are λ_{12}/λ_1 discontinuities with equal spacing λ_{12}, and within each λ_{12} there are $\lambda_{123}/\lambda_{12}$ discontinuities on $\phi_{12}(x, y)$. The absolute phase $\Phi_1(x, y)$ in terms of λ_1 can be calculated as

$$\Phi_1(x, y) = \phi_1(x, y) + 2\pi \left(\text{int} \left(\frac{\left(\frac{\lambda_{123}}{\lambda_{12}} \phi_{123}(x, y) \right)}{2\pi} \right) \times \frac{\lambda_{12}}{\lambda_1} + \text{int} \left(\frac{\left(\frac{\lambda_{12}}{\lambda_1} \phi_{12}(x, y) \right)}{2\pi} \right) \right)$$

$$= \phi_1(x, y) + 2\pi \left(\text{int} \left(\frac{\phi_{123}(x, y)}{2\pi} \times \frac{\lambda_{123}}{\lambda_{12}} \right) \times \frac{\lambda_{12}}{\lambda_1} + \text{int} \left(\frac{\phi_{12}(x, y)}{2\pi} \times \frac{\lambda_{12}}{\lambda_1} \right) \right)$$

$$(5)$$

where the function int() represents the round operation. When consider the measurement noise the four pattern images is rewritten as

$$I_k(x, y) = a(x, y) + b(x, y) \sin \left(\phi(x, y) + \frac{\pi \times k}{2} \right) + Gau(x, y), \ k = 0, 1, 2, 3 \quad (6)$$

where $Gau(x, y)$ means the white Gaussian noise item. Because that the measurement noise is propagated and even enhanced in the calculations of the equivalent phases by Eq. (4), the performance degradation in the calculations of the absolute phased with (5).

2   Method

To deal with the problems noted above we propose an improvement TWPSP method. Support there are W pixels in the direction vertical to the fringes trips, and three fringe patterns are projected with wavelengths $\lambda_1' = 1008$, $\lambda_2' = 144$ and $\lambda_3' = 16$ respectively. We simulate four pattern images with Eq. (6), and add noise to the signal. For every wavelength, the corresponding wrapped phase value $\phi_1'(x, y)$, $\phi_2'(x, y)$ and $\phi_3'(x, y)$ can be calculated directly by Eq. (3). In order to avoid Eq. (4), the three wavelengths following:

$$N_1 = \lambda_1' / \lambda_2', \ N_2 = \lambda_2' / \lambda_3' \quad (7)$$

where N_1 and N_2 are integers. As $\phi_1(x, y)$ does not have any discontinuity, there are maximally N_1 discontinuities on $\phi_2(x, y)$. Also, for each λ_2' pixels on $\phi_3(x, y)$, the same as N_2. Therefore, there are maximally $N_1 \times N_2$ discontinuities on $\phi_3(x, y)$, which can be eliminated by phase unwrapping operation in terms of phase $\Phi_3(x, y)$:

$$\Phi_3(x, y) = \phi_3'(x, y) + 2\pi \left(\text{int} \left(\frac{\phi_1'(x, y)}{\frac{2\pi}{N_1}} \right) \times N_2 + \text{int} \left(\frac{\phi_2'(x, y)}{\frac{2\pi}{N_2}} \right) \right)$$

$$= \phi_3'(x, y) + 2\pi \left(\text{int} \left(\frac{\phi_1'(x, y)}{2\pi} \times N_1 \right) \times N_2 + \text{int} \left(\frac{\phi_2'(x, y)}{2\pi} \times N_2 \right) \right)$$

$$(8)$$

Note that when $\phi_3'(x, y) = 2\pi$, a discontinuity will occur. In order to remove such a discontinuity, the phase unwrapping equation of $\Phi_3(x, y)$ should be revised to:

$$
\Phi_3(x, y) = \begin{cases}
\phi_3'(x,y) + 2\pi\left(\text{int}\left(\dfrac{\phi_1'(x,y)}{2\pi} \times N_1\right) \times N_2 + \text{int}\left(\dfrac{\phi_1'(x,y)}{2\pi} \times N_2\right)\right), \\
\quad \text{when } \phi_3'(x,y) \neq 2\pi, \ \phi_2'(x,y) \neq 2\pi \text{ and } \phi_1'(x,y) \neq 2\pi \\[4pt]
\phi_3'(x,y) + 2\pi\left(\text{int}\left(\dfrac{\phi_1'(x,y)}{2\pi} \times N_1\right) \times N_2 + \text{int}\left(\left(\dfrac{\phi_2'(x,y)}{2\pi} \times N_2\right) - 1\right)\right), \\
\quad \text{when } \phi_3'(x,y) = 2\pi \text{ or } \phi_2'(x,y) = 2\pi, \text{ and } \phi_1'(x,y) \neq 2\pi \\[4pt]
\phi_3'(x,y) + 2\pi\left(\text{int}\left(\dfrac{\phi_1'(x,y)}{2\pi} \times N_1 - 1\right) \times N_2 + \text{int}\left(\dfrac{\phi_2'(x,y)}{2\pi} \times N_2\right)\right), \\
\quad \text{when } \phi_3'(x,y) \neq 2\pi, \ \phi_2'(x,y) \neq 2\pi \text{ and } \phi_1'(x,y) = 2\pi
\end{cases}
$$

$$(9)$$

Hence the formula for unwrapping the phase map $\phi_3(x, y)$ is obtained. Although Eq. (8) is similar to Eq. (5), the wrapped phase $\phi_1'(xy)$, $\phi_2'(xy)$ can be directly calculated by Eq. (3). So the equivalent wavelengths and their phases are not need to calculate more. As well, the performance of absolute phase is revertible, and the phase unwrapping is more robust.

3   Simulation and Results

We used the fringe patterns covering 800 pixels along the vertical direction, i.e. $W = 800$. In the original TWPSP method, the wavelengths are set as $\lambda_1 = 21$, $\lambda_2 = 18$ and $\lambda_3 = 16$. For our method, we changed the three wavelengths to $\lambda_1' = 1008$, $\lambda_2' = 144$ and $\lambda_3' = 16$. The white Gaussian noises with three different Signal to Noise Ratio (SNRs) are added to the four pattern images. There covered wrapped phases of three wavelengths $\lambda = 1008$, $\lambda = 144$ and $\lambda = 16$ are demonstrated in Fig. 1. Here, SNR = 10 dB in Fig. 1, where the wrapped phases by the classical TWPSP method are also shown for comparisons. The red lines of Fig. 1a denote the wrapped phase, for the classical TWPSP method with λ_3, and the green lines of Fig. 1a represent the wrapped phase for the proposed TWPSP method with λ_3'. Because $\lambda_3 = \lambda_3'$ in the two methods, as well as the calculation of the wrapped phases are same, the wrapped phases should be same. The wrapped phase $\phi_{12}(x, y)$ is computed from the first two patterns, as shown in red lines; while the wrapped phase of λ_2' for the proposed TWPSP method is shown by the green lines. The blue lines in Fig. 1 represent the ideal wrapped phases. From the simulations results, the error propagation and noise enhancement in the classical TWSP method can be clearly observed by comparing the red curves in Fig. 1. Benefiting from the

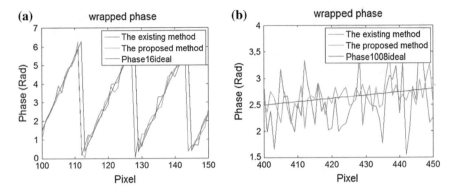

Fig. 1 The wrapped phase with the three wavelengths when SNR = 20 dB. The phase in the left sub-figure is 16 ideal, while the right sub-figure is 1008 ideal

Table 1 The root mean square errors (RMSE) of the two methods

	The ideal wrapped phase of $\lambda = 1008$	The wrapped phase of λ' in the proposed method	The wrapped phase of λ_{123} in the previous method
RMSE with SNR = 10 dB	0	0.242	0.563
RMSE with SNR = 15 dB	0	0.125	0.317
RMSE with SNR = 15 dB	0	0.064	0.176

avoidance of the calculation of the equivalent phase maps, the unwrapping results using the proposed method got significant improvement. Next, we analyze the impact of the noise quantitatively. We selected 200 points in the phase map (from 400 to 599) randomly, and compared their root mean squared errors (RMSE) in Table 1.

From the Table 1, we can see that these RMSEs of the proposed method are restrained referring to the previous method. The difference becomes larger with the decrease of the SNR, i.e., compared to the original method, the proposed method is more prominent when the measurement noise becomes larger. Moreover, experimental results are provided to verify the proposed method. In the experiment, the fringe patterns are projected onto a plaster hand model. For the classical TWPSP method in [11], the three original wavelengths are used to calculate the equivalent wavelength λ_{12}, λ_{23} and λ_{123} based on Eq. (1), yielding $\lambda_{12} = 126$, $\lambda_{23} = 144$ and $\lambda_{123} = 1008$. The wrapped phase maps from three original and equivalent wavelength of the classical TWPSP method are shown in Fig. 2. The wavelength λ'_1 in the proposed method is equal to the equivalent wavelength λ_{123}, which is used in

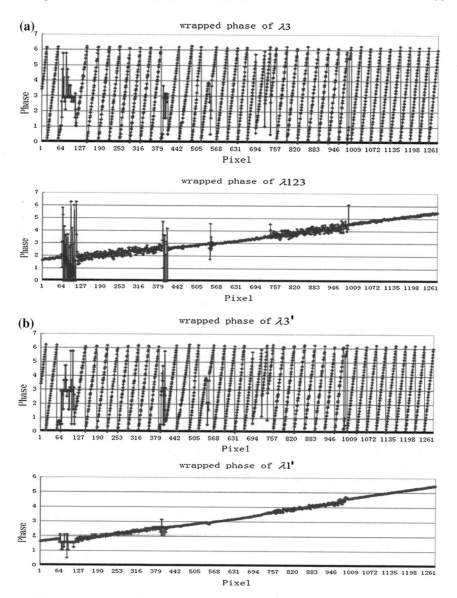

Fig. 2 The unwrapped phase of the proposed TWPSP method. **a** are the results of λ_3 and λ_{123} in existing methods. **b** are the results of λ'_3 and λ'_1 correspondingly

the classical TWPSP method. Meanwhile, the wavelength λ_2' in the proposed method equals to the equivalent wavelength λ_{23} used in classical. But λ_3' are same in both of the methods. The wrapped phase maps from the proposed TWPSP method are shown in Fig. 2.

4 Conclusion

In this paper, we proposed a novel TWPSP method. In this method, the wrapped phase can be calculated with the three specified wavelengths directly. The calculations of the equivalent wavelengths and their corresponding phase maps are not necessary yet. Moreover, the noises are greatly restrained. Since the proposed method does not need to calculate the equivalent phase maps, the implement is improved than the classical TWPSP.

References

1. Gutiérrez-García, J.C., Mosiño, J.F., Martínez, A., Gutiérrez-García, T.A., Vázquez-Domínguez, E., Arroyo-Cabrales, J.: Practical eight-frame algorithms for fringe projection profilometry. Opt. Express **21**(1), 903–917 (2013)
2. Li, S., Chen, W., Su, X.: Reliability guided phase unwrapping in wavelet transform profilometry. Appl. Opt. **473**, 369–3399 (2008)
3. Feng, S.J., Zhang, Y.Z., Chen, Q., Zuo, C., Li, R.B., Shen, G.C.: General solution for high dynamic range three-dimensional shape measurement using the fringe projection technique. Opt. Lasers Eng. **59**, 56–71 (2014)
4. Huntley, J.M., Saldner, H.O.: Error-reduction methods for shape measurement by temporal phase unwrapping, J. Opt. Soc. Am. A (14), 3188–3196 (1997)
5. Chen, Y.M., He, Y.M., Hu, E.Y., Zhu, H.M.: Deformation measurement using dual-frequency projection grating phase-shift profilometry. Acta Mech. Solida Sin. **21**(2), 110–115 (2008)
6. Towers, C.E., Towers, D.P., Jones, J.D.C.: Optimum frequency selection in multi-frequency interferometry. Opt. Lett. **28**, 1–3 (2003)
7. Towers, C.E., Towers, D.P., Jones, J.D.C.: Absolute fringe order calculation using optimized multi-frequency selection in full-field profilometry. Opt. Laser Eng. **43**, 788–800 (2005)
8. Li, Y., Xiao, W., Pan, F.: Multiple wavelength scanning based phase unwrapping method for digital holographic microscopy. Appl. Opt. **53**(5), 979–987 (2014)
9. Ding, Y., Xi, J.T., Yu, Y.G., Cheng, W.Q., Wang, S., Chicharo, J.: Frequency selection in absolute phase maps recovery with two frequency projection fringes. Opt. Express **12**, 13238–13251 (2012)
10. Long, J.L., Xi, J.T., Zhu, M., Cheng, W.Q., Cheng, R., Li, Z.W., Shi, Y.S.: Absolute phase map recovery of two fringe patterns with flexible selection of fringe wavelengths. Appl. Opt. **53**(9), 1794–1801 (2014)
11. Song, L.M., Chang, Y.L., Li, Z.Y., Wang, P.Q., Xing, G.X., Xi, J.T.: Application of global phase filtering method in multi frequency measurement. Opt. Express **22**(11), 13641–13647 (2014)
12. Lu, H., Li, Y., Serikawa, S., Li, X., Li, J., Li, K.C.: 3D underwater scene reconstruction through descattering and colour correction. **12**(4), 352–359 (2016)
13. Lu, H., Li, Y., Chen, M., Kim, H., Serikawa, S.: Brain intelligence: go beyond artificial intelligence. Mobile Netw. Appl. 1–10 (2017)
14. Lu, H., Li, Y., Mu, S., Wang, D., Kim, H., Serikawa, S.: Motor anomaly detection for unmanned aerial vehicles using reinforcement learning. IEEE Int. Things J. https://doi.org/10.1109/JIOT.2017.2737479

15. Lu, H., Li, B., Zhu, J., Li, Y., Li, Y., He, L., Li, J., Serikawa, S.: Wound intensity correction and segmentation with convolutional neural networks. Concurrency Comput. Practice Exp. https://doi.org/10.1002/cpe.3927
16. Serikawa, S., Lu, H.: Underwater image Dehazing using joint trilateral filter. Comput. Electr. Eng. **40**(1), 41–50 (2014)

The Research on the Lung Tumor Imaging Based on the Electrical Impedance Tomography

Huiquan Wang, Yanbo Feng, Jinhai Wang, Haofeng Qi, Zhe Zhao
and Ruijuan Chen

Abstract Magnetic detection electrical impedance tomography (MD-EIT) can be used to reconstruct images of conductivity from magnetic field measurement taken around the body in vivo. To achieve the lung tumor imaging based on the MD-EIT, this study established the human thoracic model according to CT image. The Moore-Penrose inverse, TSVD and Tikhonov regularization were tried before the reconstruction to improve imaging accuracy respectively. Four levels of noise were added into the simulation data to meet the real detection situation (the SNR is 30, 60, 90 and 120 dB). The reconstruction results, which were pre-processed by TSVD regularization, had the best performance. The average relative error (ARE) values of current density distribution equals 0.21 and 0.22 for healthy and lung tumor person respectively. The tumor in lung can be distinguished clearly from the MD-EIT image. The MD-EIT is one of the most promising technology in dynamic lung imaging for clinical application.

Keywords MD-EIT · Lung tumor · Forward problem · Inverse problem

H. Wang · Y. Feng · J. Wang · H. Qi · Z. Zhao (✉) · R. Chen (✉)
School of Electronics and Information Engineering, Tianjin Polytechnic University,
399 Binshui West Street, Xiqing District, Tianjin 300387, People's Republic of China
e-mail: zhaozhe@tjpu.edu.cn

R. Chen
e-mail: chenruijuan@tjpu.edu.cn

Z. Zhao
Tianjin Key Laboratory of Optoelectronic Detection Technology and System,
399 Binshui West Street, Xiqing District, Tianjin 300387, People's Republic of China

H. Wang · J. Wang · R. Chen
Tianjin Medical Electronic Treating-Technology Engineering Center, Tianjin 300387,
People's Republic of China

© Springer International Publishing AG 2018
H. Lu and X. Xu (eds.), *Artificial Intelligence and Robotics*,
Studies in Computational Intelligence 752,
https://doi.org/10.1007/978-3-319-69877-9_11

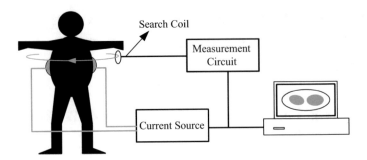

Fig. 1 Magnetic detection electrical impedance tomography system

1 Introduction

According to the world cancer report [1] of World Health Organization, the lung tumor is the most dangerous disease for human. The electrical resistivity of mammalian tissues varies widely and is correlated with physiological function [2]. MD-EIT is an experimental technique that aims to produce images of conductivity from magnetic field measurements taken around the body [3, 4] and is closely related to the medical imaging technique of electrical impedance tomography (EIT) [5], as shown in Fig. 1 is the MD-EIT system.

Coil is used to measure the magnetic filed in free-space instead of electrode in EIT, which is a effectual approach to eliminate contact impedance and provide enough magnetic filed data [6–8]. Unlike the inverse problem of EIT, in MD-EIT the relationship between current density and induced magnetic flux density is linear, though the subsequent reconstruction of conductivity from current density is non-linear, it is in general reasonably well posed [9].

2 Method and Results

2.1 Experiment on General Model

A general three-dimensional model is constructed, as shown in Fig. 2a, to simulate the structure of human chest. The size of cube is 10 m × 10 m × 10 m, and cylinders with radius of 1 m are put at both sides of the mid line on bottom symmetrically. The electrical conductivity of two cylinders is set as 1 S/m and the rest area is set as 0.548 S/m. Two electrodes are attached on the surfaces perpendicular to X axis with the current intensity of 1 A. The finite element method is applied to the forward problem, as shown in Fig. 2b is the meshing result. The model is divided into elements with size of 1 m × 1 m × 1 m, so the total number of cube elements is 1000, for the nodes is 1331.

The current density distribution of the middle cross-section perpendicular to Z axis of the model is calculated as shown in Fig. 3a. Based on the data of magnetic

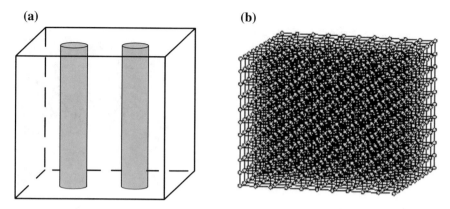

Fig. 2 Cubic model (**a**) and meshing result (**b**)

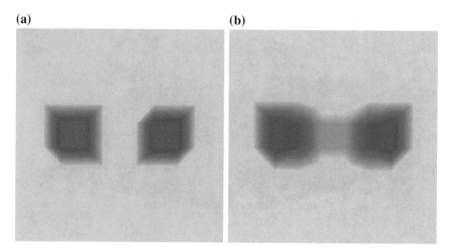

Fig. 3 Current density distribution. **a** is from forward problem, **b** is reconstructed image

flux density in forward problem, the Moore-Penrose inverse is used to reconstruct the current density distribution, the result is shown as Fig. 3b.

2.2 Thoracic Models

As shown in Fig. 4 are the CT images, which are from the Lung Image Database Consortium (LIDC) [10], Fig. 4a is of healthy person, Fig. 4b is of person with lung tumor. To obtain organ profile and the spatial structure of chest, the image segmentation based on gray value is applied to the CT image [11–15]. Afterward, the three-dimensional chest model, closely similar to real chest, is constructed as shown in Fig. 5.

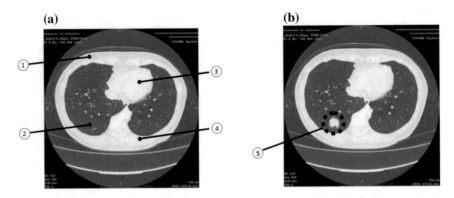

Fig. 4 Computed tomography images. it can be recognized in (**a**) that 1 is muscle, 2 is lung, 3 is heart, 4 is bone tissue. In (**b**) 5 is lung tumor. It should be noticed that both two images are from one healthy person, the lung tumor tissue (**b**) 0.5 is from another CT image of actual patient, who is with same age and morphological characteristics to the healthy person

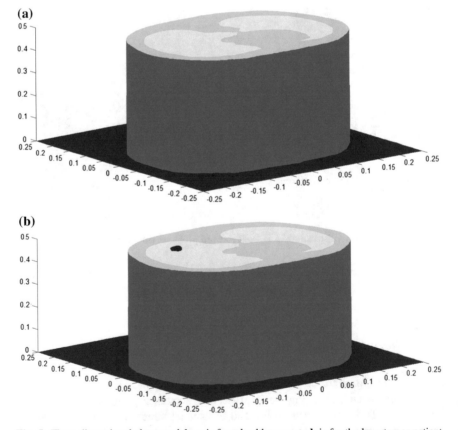

Fig. 5 Three-dimensional chest models. **a** is from healthy person, **b** is for the lung tumor patient. Every pixel is set to 1 mm, the maximum length and width of chest model is 43 and 21 cm, height is 50 cm, the minimum radius of tumor is 3 cm. The conductivity of lung area and tumor is set to 0.04 and 1.45 S/m, the other is set as muscle to 0.58 S/m [16]

2.3   Results in Forward Problem and Inverse Problem

The finite element method is applied to calculate the forward problem, as shown in Fig. 6. Firstly, electrical potential distribution is calculated, and then current density distribution is solved as shown in Fig. 7. For the inverse problem, 3600 points of magnetic induction intensity data is calculated in forward problem. Moore-Penrose inverse is used to reconstruct the current density distribution images, the results are shown in Fig. 8. In order to meet the actual measurements, we add noise to calculated data with the SNR of 30, 60, 90 and 120 dB, using Moore-Penrose inverse, Tikhonov and TSVD regularization to reconstruct the image of current density distribution respectively, as shown in Fig. 9.

In order to assess the quality of reconstructed current density, average relative errors (ARE) is introduced to do assessment,

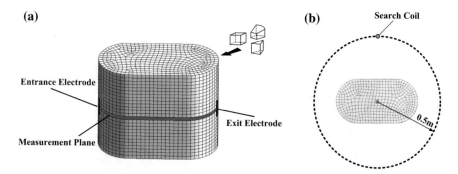

Fig. 6 Three-dimensional mesh, electrode position and measurement plan. As **a** shown, the chest model is divided into 18820 elements with 21087 nodes. Two rectangular electrodes with 1 cm length and 1 cm width, regardless the thickness. The current with frequency of 31.25 kHz and amplify of 50 mA is applied to the model through electrodes. **b** shows the measurement plane, which is at middle of the model, around the circle with radius of 0.5 m the magnetic field is detected

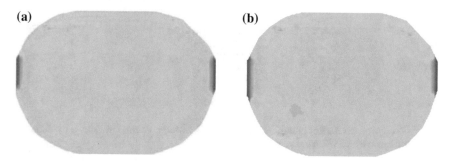

Fig. 7 The current density distribution. **a** is of healthy person, **b** is of lung tumor patient

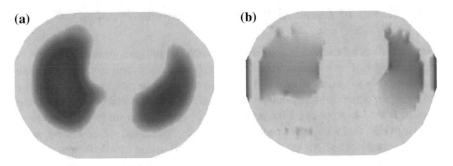

Fig. 8 Reconstruction of the current density distribution. **a** is of healthy person, **b** is of lung tumor patient

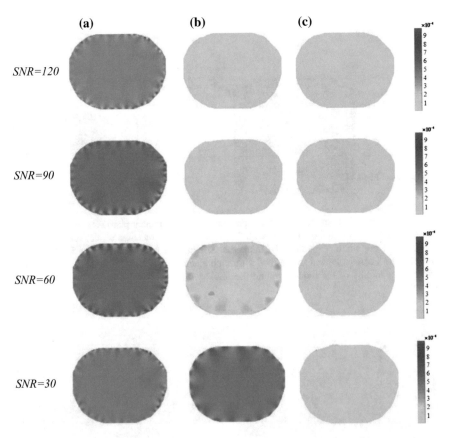

Fig. 9 Reconstructed images in noise environment. **a** by Moore-Penrose inverse. **b** by Tikhonov regularization. **c** by TSVD regularization

Fig. 10 The comparison of reconstruction quality using three reconstruction method

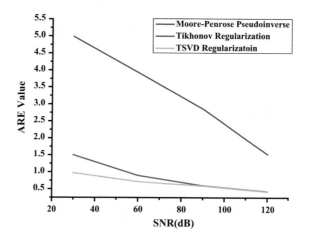

$$\mathrm{ARE} = \frac{\|J - j\|_2}{\|J\|_2} \qquad (10)$$

where J is original current density, j is reconstructed current density. ARE reflects the precision of j related to J. In noise-free environment, ARE of healthy person and lung tumor patient equals 0.21 and 0.22 respectively. For the results of three reconstruction method in four noise level, ARE is also used to do the assessment objectively, as shown in Fig. 10.

3 Discussion and Conclusion

The reconstruction results have shown that the accuracy of measured data directly affects the quality of reconstructed images, regularization must be applied to reconstruction. In noise-free environment, the shape of lung and location of tumor can be easily recognized in the images reconstructed by Moore-Penrose inverse as shown in Fig. 8. As long as noise added, regularization must be used to reduce the influence of noise as shown in Fig. 9. Tikhonov and TSVD regularization are applied to the reconstruction. TSVD regularization has better reconstruction result as shown in Fig. 10. The inverse problem of MD-EIT is a kind of ill-condition problem without an unique solution. The regularization reduces the effects of high-frequency components of the data, for instance noise and edged, which makes regularized solutions tend to be smoother.

Those results provide the foundation for MD-EIT to be used in detecting and monitoring lung tumors. From the reconstructed images, lung tissue and tumor can be recognized. Different from modern popular imaging system, with the character of noncontact, portability, dynamic imaging, functional imaging and relatively inexpensive, MD-EIT is a strong complement to existing technology.

Acknowledgements We thank Dr. Guangxu Li, Dr. Yu Zheng, Lei Tian their help in field measurements and programming. This study financially supported by the 61th Batch of China Postdoctoral Science Foundation and Key Laboratory of Biomedical Effects of Nanomaterials and Nanosafety, CASCNSKF 2016027.

References

1. Stewart, B., Wild, C.: World cancer report 2014. In: International Agency for Research on Cancer (2014)
2. Metherall, P., Barber, D.C., Smallwood, R.H., et al.: Three-dimensional electrical impedance tomography. Nature **380**(6574), 509 (1996)
3. Ireland, R.H.: Anatomically Constrained Image Reconstruction Applied to Emission Computed Tomography and Magnetic Impedance Tomography. University of Sheffield (2002)
4. Tozer, J.C., Ireland, R.H., Barber, D.C., et al.: Magnetic impedance tomography. Ann. N.Y. Acad. Sci. **873**(1), 353–359 (1999)
5. Barber, D.C., Brown, B.H.: Review article: applied potential tomography. J. Phys. E: Sci. Instrum. **17**(9), 723–733 (1984)
6. Brown, B.H.: Electrical impedance tomography (EIT): a review. J. Med. Eng. Technol. **27**(3), 97 (2003)
7. Costa, E.L.V., Amato, M.B.P.: Electrical impedance tomography in critically ill patients. Clin. Pulm Med. **20**, 178–186 (2013)
8. Kolehmainen, V., Vauhkonen, M., Karjalainen, P.A., Kaipio, J.P.: Assessment of errors in static electrical impedance tomography with adjacent and trigonometric current patterns. Physiol. Meas. **18**, 289–303 (1997)
9. Ireland, R.H., Tozer, J.C., Barker, A.T., et al.: Towards magnetic detection electrical impedance tomography: data acquisition and image reconstruction of current density in phantoms and in vivo. Physiol. Meas. **25**(3), 775 (2004)
10. Armato, S.G., Mclennan, G., Bidaut, L., et al.: The lung image database consortium (LIDC) and image database resource initiative (IDRI): a completed reference database of lung nodules on CT scans. Med. Phys. **38**(2), 915–931 (2011)
11. Yoshino, Y., Miyajima, T., Lu, H., et al.: Automatic classification of lung nodules on MDCT images with the temporal subtraction technique. Int. J. Comput. Assist. Radiol. Surg. 1–10 (2017)
12. Lu, H., Li, B., Zhu, J., et al.: Wound intensity correction and segmentation with convolutional neural networks. Concurrency Comput. Pract. Experience (2016)
13. Lu, H., Li, Y., Chen, M., Kim, H., Serikawa, S.: Brain intelligence: go beyond artificial intelligence. Mob. Netw. Appl. 1–10 (2017)
14. Lu, H., Li, Y., Mu, S., Wang, D., Kim, H., Serikawa, S.: Motor anomaly detection for unmanned aerial vehicles using reinforcement learning. IEEE Internet Things J. https://doi.org/10.1109/JIOT.2017.2737479
15. Serikawa, S., Lu, H.: Underwater image dehazing using joint trilateral filter. Comput. Electr. Eng. **40**(1), 41–50 (2014)
16. Faes, T.J., Ha, V.D.M., de Munck, J.C., et al.: The electric resistivity of human tissues (100 Hz–10 MHz): a meta-analysis of review studies. Physiol. Meas. **20**(4), R1 (1999)

Combining CNN and MRF for Road Detection

Lei Geng, Jiangdong Sun, Zhitao Xiao, Fang Zhang and Jun Wu

Abstract Road detection aims at detecting the (drivable) road surface ahead vehicle and plays a crucial role in driver assistance system. To improve the accuracy and robustness of road detection approaches in complex environments, a new road detection method based on CNN (convolutional neural network) and MRF (markov random field) is proposed. The original road image is segmented into super-pixels of uniform size using simple linear iterative clustering (SLIC) algorithm. On this basis, we train the CNN which can automatically learn the features that are most beneficial to classification. Then, the trained CNN is applied to classify road region and non-road region. Finally, based on the relationship between the super-pixels neighborhood, we utilize MRF to optimize the classification results of CNN. Quantitative and qualitative experiments on the publicly datasets demonstrate that the proposed method is robust in complex environments. Furthermore, compared with state-of-the-art algorithms, the approach provides the better performance.

Keywords Driver assistance system · Road detection · Super-pixel CNN · MRF

L. Geng · J. Sun · Z. Xiao (✉) · F. Zhang · J. Wu
School of Electronics and Information Engineering,
Tianjin Polytechnic University, Tianjin, China
e-mail: xiaozhitao@tjpu.edu.cn

L. Geng
e-mail: genglei@tjpu.edu.cn

J. Sun
e-mail: sjdakm@163.com

F. Zhang
e-mail: zhangfang@tjpu.edu.cn

J. Wu
e-mail: wujun@tjpu.edu.cn

L. Geng · J. Sun · Z. Xiao · F. Zhang · J. Wu
Tianjin Key Laboratory of Optoelectronic Detection Technology
and Systems, Tianjin, China

© Springer International Publishing AG 2018
H. Lu and X. Xu (eds.), *Artificial Intelligence and Robotics*,
Studies in Computational Intelligence 752,
https://doi.org/10.1007/978-3-319-69877-9_12

103

1 Introduction

Research shows that about 1.4 million people dead in road traffic accident at home and abroad every year, owing to the driver fatigue and inattention [1]. Driver assistance system can reduce the incidence of traffic accidents by reminding and guiding the driver [2]. Road detection based on vision is the key to driver assistance system which can not only provide clues for obstacle detection but also help path planning [2].

At present, road detection methods are divided into the following two categories: the first method is based on model. For instance, Wang et al. [3] converted the image into an inverse perspective space and preprocessed the image using a hybrid Gaussian anisotropic filter, then used the Bezier spline curve to construct variable road template. Finally they realized the road detection by using the improved RANSAC algorithm to solve the template parameter. Wang et al. [4] combined parabolic model and Hough transform to detect road boundary. They first utilized Hough transform to fit edge points and then used parabolic model to fit curved road boundary. Kong et al. [5] utilized multi-scale Gabor wavelet transform to calculate the texture direction of pixels and estimated the vanishing point of road by adaptive voting. Finally, they utilized vanishing point to constraint edge detection and then got road boundary. Such methods can detect structured road accurately and the detection result relatively complete. However, with the movement of vehicles in reality, the shape of the road is constantly changing. To establish a robust model for road matching is a challenge problem. One of the most common method is the second category, which based on feature. For instance, Alvarez and ĽOpez [6] paid attention to road detection under different lighting conditions. They used camera parameters to restore the illuminant-invariant feature space and then did the road modeling according to road grayscale value. Pixels were labeled by combining a model–based classifier in the last. Mendes et al. [7] proposed a network to detect road that takes advantage of a large contextual window and uses a Network-in-Network (NiN) proposed a multi-layer CNN architecture with a new loss function to detect free space. Brostow et al. [8] integrated appearance features and structure cues which got from 3D point clouds, and then used a randomized decision forest to get road regions. A framework proposed by Sturgess et al. [9] combined motion and appearance features. A conditional random field (CRF) was then employed to integrate higher order potentials and such features. By minimizing the energy function, the labels were got. Yuan et al. [10] extracted HOG, LBP, DSIFT features of the area around the road border to train SSVM classifier, and then get a series of sparse points that were the candidates of the road boundaries. Finally, they fitted the boundary by using RANSAC algorithm. Fernández et al. [11] made use of the watershed transformation to segment the original image into super-pixels and then extracted color, texture features of the super-pixels to train decision trees based classifiers to complete road detection. Alvarez et al. [12] combined CNN with a color space-based texture descriptor, labeled pixels by comparing multiplying probability and the threshold. Such methods mainly use color, texture, edge features and so on to detect road, it is not sensitive to the shape of the road and can be suitable for any shape of the road. However, it is easily influenced by shadows or lights when the road scene is complex and changeable which makes more difficult to choose features.

In summary, to improve the robustness of road detection method in complex environments, we propose a new approach to integrate CNN and MRF for road detection. The main contributions made in this paper are: (i) Design a novel network structure that is suitable for road detection. (ii) Conduct label optimize with a MRF model to improve the accuracy of detection. The overall accuracy of our method on the Cambridge-driving Labeled Video Database (CamVid) [13] can reach 96.1%. The outline of the paper is organized as follows: In Sect. 2 we discuss the main part of the proposed method. Section 3 describes the details of the database and the network structure. Different experiments on the database are given. Finally, conclusions and directions for future work are drawn in Sect. 4.

2 Road Detection

2.1 Super-Pixel Extraction

Super-pixel can extract the local features, obtain more efficient structural information and reduce the computational complexity of the subsequent processing greatly than pixel which is the basic unit in the traditional processing method. Compared with fixed-size rectangle-based window, a single super-pixel usually describes only one target, avoiding confounding targets in rectangle-based window. Compared with other super-pixel segmentation algorithm such as Normalized-Cuts, Graph-Cuts, Turbo-Pixel, Quick-Shift and Mean-Shift [14], SLIC algorithm has the advantages of regular region shape by pre-segmentation, less time-consuming and can preserve the target boundary better.

SLIC algorithm is a super-pixel segmentation method based on K-means clustering, using the color similarity and the positional relationship of the pixels to generate super-pixels [15]. Different super-pixel segmentation results of the road image are illustrated in Fig. 1.

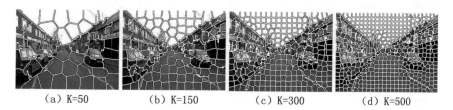

(a) K=50   (b) K=150   (c) K=300   (d) K=500

Fig. 1 The segmentation results of different super-pixel numbers

Results are shown in Fig. 1, the image size is 320 × 240 and (a)–(d) correspond to different super-pixel numbers respectively. We can find that when the number of super-pixel is little, some super-pixels will contain both the road region and non-road region. When the number of super-pixel is equal to 300, a super-pixel contains only one region which can meet the needs of the subsequent classification. Nevertheless, when the number of super-pixel is equal to 500, the edges remain well, but the more the number of super-pixel is, the larger the amount of computation is. At the same time, it is not conducive to extract super-pixel features and will affect the following classification because one super-pixel contains too few pixels. Therefore, we segment the image into 300 super-pixels.

2.2   CNN

Due to the nature of the scene, good feature extraction and feature expression play a very important role in the road detection. Traditional classifiers need to design features for each situation artificially, and it takes a lot of time to verify whether these features can distinguish the road. CNN has an intelligent learning mechanism, which transforms the feature expression of the sample in the original space into a new feature space through feature transform layer by layer. It makes classify or predict easier, and shows a powerful ability to learn the essential features of a data set from a small number of samples. Meanwhile, it is widely used in the field of object detection [16], object recognition [17], segmentation [18] and other fields [19–22]. The common CNN is mainly composed of convolution layer (C-Layer), down-sampling layer (S-Layer) and fully connected layer (F-Layer), as shown in Fig. 2.

CNN utilize the original image as the input for the input layer. Each neuron in hidden layer takes the input layer or the output of adjacent neuron in previous hidden layer as input. In the process of the CNN learning, the current network

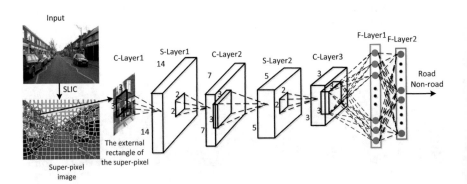

Fig. 2 CNN structure

weights and the network inputs are used to compute the network outputs. Then the outputs and the sample labels are made use of to calculate error. The back propagation algorithm is utilized to calculate the derivative of error to network weights. In the end, the weights are updated by using the weight updating method. After several times of training, the parameters of the network can be optimized and a stable network structure is obtained.

The training processes are explained as follows:

- The SLIC algorithm is used to extract the super-pixels from the original image;
- Centering on the center of the super-pixel, the external rectangle of the super-pixel is preserved;
- The rectangle is normalized to the $N \times N$ size and then labeled (road is positive sample, labeled 1. non-road is negative sample, labeled 0). All rectangles and its corresponding label are used as training data;
- CNN utilize the training data as input to train the network.

The detection processes are as follows: We segment the road image into super-pixels. The external rectangle of the super-pixel is extracted and resized to $N \times N$ size. Then through the trained CNN determine whether it is the road, the initial road region is obtained.

2.3 MRF for Label Optimization

In complex environments, the accuracy of classification is difficult to reach 100%. The labelling result obtained by the CNN is only a rough estimation. To get more accurate results, we utilize the MRF to optimize the label map. Simple MRF refers to that the current state of one thing is only related to its previous one or n state but not related to the earlier state before it. Extended to the field of image, it is considered that the features of a certain point in the image (for instance gray value, color value, etc.) are only related to a small area near it and independent of other fields.

Let x and y be the random field in the two-dimensional plane, y is observed image, x denotes label field, $\hat{x}$ is an estimate of the true label of the image. According to the maximum a posteriori criterion (MAP), the formula (1) is obtained.

$$\hat{x} = \arg\max\{P(y/x)P(x)\} \tag{1}$$

where $P(x)$ is called the prior probability of the label field, can be described as Gibbs distribution [23] and formulated as follows:

$$P(x_i) = \frac{\exp(-\mu(x_i))}{\sum_{x_i \in L} \exp(-\mu(x_i))} \tag{2}$$

$$\mu(x_i) = -\beta_i \sum_{i \in N_i} [\sigma(x_i, y_i) - 1] \tag{3}$$

$$\sigma(x_i, y_i) = \begin{cases} 1 & x_i = y_i \\ 0 & x_i \neq y_i \end{cases} \tag{4}$$

where $P(y/x)$ denotes likelihood probability, and follows the formula (5). When the label is given $x_i = l$, it is generally considered that the pixel intensity value obeys the Gauss distribution with the parameter of $\theta_i = \{\mu_l, \sigma_l\}$.

$$P(y_i/x_i) = \left(1/\sqrt{2\pi\sigma_l^2}\right) \cdot \exp\left(-(y_i - \mu_l)^2/2\sigma_l^2\right) \tag{5}$$

where x_i is the label of super-pixel i and y_i is the average pixel value of all the pixels in the super-pixel. The value of l is 1 or 0, super-pixel represents road when $l = 1$, otherwise represents non-road. μ_l stands for the mean of the l region and σ_l stands for the standard variance of the l region. N_i is the neighborhood of super-pixel i. β is a smoothing parameter (ranges from 0.8 to 1.4).

Algorithm steps are as follows:

- The result of CNN classification is used as the initial segmentation of the image;
- $\theta_i = \{\mu_l, \sigma_l\}$ is updated by the current segmentation, μ_l and σ_l are the mean and standard variance of the current l region;
- Calculate the maximum possible category of each super-pixel according to the current image parameters, the last segmentation results and the formula (1);
- If the procedure is convergent or reach the maximum number of iterations, then exit, otherwise return to step 2 and do the next iteration.

3 Experiments

3.1 Experimental Dataset

The proposed method is achieved based on VS2012 compiler environment, and the experimental test platform is Intel (R) Core (TM) I3-350 M core processor, 2.27 GHz frequency, 3 GB memory, Windows 7 Professional 32 operating system. The experiments are performed on the CamVid containing 701 road images and their corresponding Ground Truth (GT) images. The original image size is 960 × 720 and the GT image resolution is 320 × 240. All these images are taken from different backgrounds, different road types and different lighting conditions, meanwhile, influenced by shadows and other vehicles in different traffic conditions.

3.2 Parameter Setup and Network Structure Design

The original image in the database was down-sampled at 320×240. According to the method in [24], 367 images were selected as the training set, 233 images were used as testing set. We follow the experimental setup in Sect. 2.2 to get the samples, 144160 training samples and 16000 test samples are obtained. Then we normalize samples to 16×16 size. In this paper, a variety of network structures are tested in the experiment, we choice RD-CNN (Road Detection-Convolution Neural Network) at the last. RD-CNN includes three C-layers, two S-layers, and two F-layers. The convolution kernel size is 3×3, and the three C-layers respectively use 16, 32, 64 convolution kernels to extract the image features. Maximum sampling is used in the two S-layers. The number of neurons in the first F-layer is 256 and the number in the second F-layer equal 2. Different parameters of different network structures are shown in Table 1. 2C-0S-16-32 has no S-layer so the recognition effect is not ideal. 2C-2S-16-32 increases two S-layers based on 2C-0S-16-32, and then the recognition rate is obviously improved; 3C-2S-4-8-16 transforms the convolution kernel of 5×5 size into two 3×3 convolution kernels, the depth of the network is increased. The recognition rate has improved, but is not obvious. On this basis, RD-CNN (3C-2S-16-32-64) increases the number of convolution kernels, and the recognition rate is further improved to 93.6%.

To compare the influence of different network structure parameters on recognition rate intuitively, Fig. 3 shows the recognition rate for each network with 400,000 iterations. It can be seen from Fig. 3 that the recognition rate of the network fluctuates greatly when the number of iterations is small. With the increase of the number of iterations, the tunable parameters of the network are gradually optimized; the recognition rate is gradually improved and tends to be stable. By increasing the S-layer, we can reduce the dimension of the output features of the C-layer, reduce information redundancy and improve network efficiency. Increasing the number of C-layers or the number of convolutional kernels in a certain range can improve the recognition rate, but also increase the amount of calculation of the network. All things considered, we choose the RD-CNN network.

Table 1 Different parameters of different network structure

Network structure	C1	S1	C2	S2	C3	F1	F2
2C-0S-16-32	$16 * 5^2$	–	$32 * 3^2$	–	–	256	2
2C-2S-16-32	$16 * 5^2$	$2 * 2$	$32 * 3^2$	$2 * 2$	–	256	2
3C-2S-4-8-16	$4 * 3^2$	$2 * 2$	$8 * 3^2$	$2 * 2$	$8 * 3^2$	256	2
3C-2S-16-32-64 (RD-CNN)	$16 * 3^2$	$2 * 2$	$32 * 3^2$	$2 * 2$	$64 * 3^2$	256	2
3C-2S-48-64-128	$48 * 3^2$	$2 * 2$	$64 * 3^2$	$2 * 2$	$128 * 3^2$	256	2

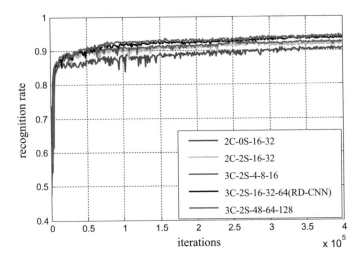

Fig. 3 The recognition rate of different networks

3.3 Experimental Results and Analysis

To verify the robustness of the proposed algorithm, we select the road detection method in [5, 7–9, 10] to be compared with it on the testing set in the database. The vanishing point based method (VP) proposed in [5] does not need any training data. It can prove the superiority of utilizing road training samples. Qualitative road detection results are shown in Fig. 4.

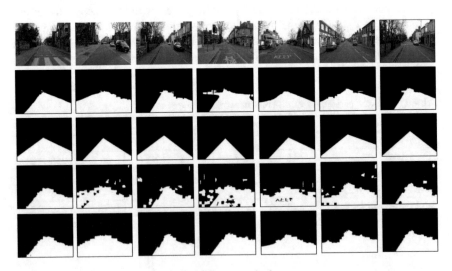

Fig. 4 Typical road detection results for different methods

Table 2 Quantitative results
of different methods

Methods	P	R	F	A
VP [5]	74.6	91.7	82.3	81.7
FCN-LC [7]	92.8	91.6	92.2	93.9
Mot. & App. [8]	92.7	86.0	89.2	89.5
App. & SFM. [9]	97.1	93.2	95.1	95.3
CPF&CNN-7 [10]	94.4	95.9	95.1	95.5
Our approach	97.2	93.6	95.5	96.1

As can be seen from Fig. 4, first row show the input image. Second row show the GT image. Third row show the road results using VP. The road likelihood using CNN are show in fourth row. In the last line of Fig. 4 show the output of the proposed algorithm using CNN and MRF. In order to show that our method can perform better than such methods, we use four evaluation indexes of recall (R), Precision (P), F-measure, accuracy (A). R reflects the ability to reveal the information you need. P indicates the correctness of the detected results. F-measure is a balance between them. A indicates the overlapping ratio between GT image and the detected result.

The evaluation result is shown in Table 2. From Fig. 4 and Table 2 we can see that our method outperforms those state-of-the-art algorithms obviously. To be specific, VP method utilize linear model to detect road, it can get good results in simple structured road, but when facing curved road or there are obstacles in the road ahead, missed and false detection is very serious. In terms of Mot. & App. [8] and App. & SFM. [9], they utilized traditional classifier and information such as appearance-based features or structure cues from motion. Those methods depend on human–designed features and are susceptible to scene changes. It can be proved the superiority and rationality of extracting the features automatically by CNN. As for FCN-LC, it is unable to correctly classify different types of road surfaces or regions subject to extreme lighting conditions. Besides, CPF&CNN-7 can adapt to different shapes of the road. But due to the effects of lights, shadows, and complex obstacles, the accuracy of any classifier is difficult to achieve 100%, and then too many road regions are divided into non-road regions. Compared with CPF&CNN-7, we can prove the superiority of using MRF.

4 Conclusions

In this work we propose an efficient road detection method based on CNN and MRF. To reduce the computational complexity and guarantee that road and non-road regions will not be mixed up, we first segmented the original image into super-pixels. For road detection, a RD-CNN is designed, which can learn the most suitable features from a large number of samples automatically and has good robustness under lighting variations and road shadows. In the end, to improve the detection accuracy, according to the relationship between the super-pixel neighborhoods, MRF model is used to optimize the detected results of RD-CNN. Quantitative and qualitative experiments on the challenging CamVid database verify that the proposed method is effective at detecting road regions. However, this database does not include harsh weather conditions. We plan to enhance the adaptability of the algorithm to bad weather in the future.

Acknowledgements This work is supported by National Natural Science Foundation of China under grant No. 61601325, the key technologies R & D program of Tianjin under grant No.14ZCZDGX00033, and research project for application foundation and frontier technology of Tianjin under grant No. 14JCYBJC42300.

References

1. Research Institute of Highway Ministry of Transport: The Blue Book of Road Safety in China 2014, pp. 13–14. China Communications Press, Beijing (2015)
2. Cao, X., Lin, R., Yan, P., Li, X.: Visual attention accelerated vehicle detection in low-altitude airborne video of urban environment. IEEE Trans. Circuits Syst. Video Technol. **22**(3), 366–378 (2012)
3. Wang, K., Huang, Z.H., Zhong, Z.H.: Algorithm for urban road detection based on uncertain bezier deformable template. J. Mech. Eng. **49**(8), 143–150 (2013)
4. Wang, J., Gu, F., Zhang, C., Zhang, G.: Lane boundary detection based on parabola model. In: IEEE International Conference on Information and Automation, pp. 1729–1734. IEEE, Harbin (2010)
5. Kong, H., Audibert, J.Y., Ponce, J.: Vanishing point detection for road detection. In: 2009 IEEE Conference on Computer Vision and Pattern Recognition (CVPR), pp. 96–103. IEEE (2009)
6. Alvarez, J.M., LOpez, A.M.: Road detection based on illuminant invariance. IEEE Trans. Intell. Transp. Syst. **12**(1), 184–193 (2011)
7. Mendes, C.C.T., Frémont, V., Wolf, D.F.: Exploiting fully convolutional neural networks for fast road detection. In: IEEE International Conference on Robotics and Automation. IEEE (2016)
8. Brostow, G.J., Shotton, J., Fauqueur, J., Cipolla, R.: Segmentation and recognition using structure from motion point clouds. In: ECCV, vol. 1, pp. 44–57 (2008)
9. Sturgess, P., Alahari, K., Ladicky, L., Torr, P.H.S.: Combining appearance and structure from motion features for road scene understanding. In: BMVC'09 (2009)
10. Yuan, Y., Jiang, Z., Wang, Q.: Video-based road detection via online structural learning. Neurocomputing **168**(C), 336–347 (2015)
11. Fernández, C., Izquierdo, R., Llorca, D.F., Sotelo, M.A: A comparative analysis of decision trees based classifiers for road detection in urban environments. In: Proceedings of the IEEE International Conference on Intelligent Transportation Systems 2015, pp. 719–724 (2015)
12. Alvarez, J.M., Gevers, T., Lecun, Y., Lopez, A.M.: Road scene segmentation from a single image. In: Computer Vision–ECCV, pp. 376–389. Springer, Berlin (2012)
13. Brostow, G.J., Fauqueur, J., Cipolla, R.: Semantic object classes in video: a high definition ground truth database. Pattern Recogn. Lett. **30**(2), 88–97 (2009)
14. Achanta, R., Shaji, A., Smith, K.: SLIC super-pixels compared to state-of-the-art super-pixels methods. IEEE Trans. Pattern Anal. Mach. Intell. **34**(11), 2274–2282 (2012)
15. Achanta, R., Shaji, A., Smith, K.: SLIC Super-pixels. Swiss federal Institute of Technology, Lausanne, Vaud, Switzerland (2010)

16. Cecotti, H., Graser, A.: Convolutional neural networks for P300 detection with application to brain-computer interfaces. IEEE Trans. Pattern Anal. Mach. Intell. **33**(3), 433–445 (2010)
17. Ji, S., Xu, W., Yang, M., et al.: 3D convolutional neural networks for human action recognition. IEEE Trans. Pattern Anal. Mach. Intell. **35**(1), 221–231 (2013)
18. Turaga, S.C., Murray, J.F., Jain, V., Roth, F., Helmstaedter, M., Briggman, K., Denk, W., Seung, H.S.: Convolutional networks can learn to generate affinity graphs for image segmentation. Neural Comp. **22**, 511–538 (2010)
19. Lu, H., Li, Y., Chen, M., Kim, H., Serikawa, S.: Brain intelligence: go beyond artificial intelligence. Mob. Netw. Appl. 1–10 (2017)
20. Lu, H., Li, Y., Mu, S., Wang, D., Kim, H., Serikawa, S.: Motor anomaly detection for unmanned aerial vehicles using reinforcement learning. IEEE Int. Things J. https://doi.org/10.1109/JIOT.2017.2737479
21. Lu, H., Li, B., Zhu, J., Li, Y., Li, Y., He, L., Li, J., Serikawa, S.: Wound intensity correction and segmentation with convolutional neural networks. Concurrency Comput. Pract. Experience. https://doi.org/10.1002/cpe.3927
22. Serikawa, S., Lu, H.: Underwater image dehazing using joint trilateral filter. Comput. Electr. Eng. **40**(1), 41–50 (2014)
23. Li, S.Z.: Markov Random Field Modeling in Computer Vision. Springer, Tokyo, Japan (1995)
24. Brostow, G.J., Fauqueur, J., Cipolla, R.: Semantic object classes in video: a high definition ground truth database. Pattern Recog. Lett. (2008)

The Increasing of Discrimination Accuracy of Waxed Apples Based on Hyperspectral Imaging Optimized by Spectral Correlation Analysis

Huiquan Wang, Haojie Zhu, Zhe Zhao, Yanfeng Zhao and Jinhai Wang

Abstract To increase the classification accuracy and stability of the prediction model, an approach to evaluate the quality of samples' hyperspectral image is needed. The spectral correlation analysis of each pixel was used to determine quality of the sample's hyperspectral image in this study. 400 hyperspectral image ROIs were extracted from 20 apples (10 apples with waxed and the other 10 apples without any waxed) and the data were separated into 300 as train set and 100 as test set randomly. The experimental group data were evaluated by the spectral correlation analysis, and only qualified data were used for model training. The control group data were all used for modeling training. The least squares support vector machine (LS-SVM) model were used to establish the classification model between the hyperspectral image and waxed situation. The prediction result showed the classification accuracy were 94% and 86% when the low-quality sample data for training were filtered by spectral correlation analysis. By evaluating the quality of the hyperspectral image measured, more reliable prediction results can be obtained, which can make the noninvasive discrimination of food safety come to the practice application sooner.

Keywords Hyperspectral image · Correlation analysis · Classification Waxed apples

H. Wang · H. Zhu · Z. Zhao (✉) · Y. Zhao · J. Wang (✉)
School of Electronics and Information Engineering, Tianjin Polytechnic University,
399 Binshui West Street, Xiqing District, Tianjin 300387, China
e-mail: zhaozhe@tjpu.edu

J. Wang
e-mail: tjpubme@126.com

Z. Zhao
Tianjin Key Laboratory of Optoelectronic Detection Technology and System,
399 Binshui West Street, Xiqing District, Tianjin 300387, People's Republic of China

H. Wang · J. Wang
Tianjin Medical Electronic Treating-Technology Engineering Center, Tianjin 300387,
People's Republic of China

© Springer International Publishing AG 2018
H. Lu and X. Xu (eds.), *Artificial Intelligence and Robotics*,
Studies in Computational Intelligence 752,
https://doi.org/10.1007/978-3-319-69877-9_13

1 Introduction

In order to keep the apples freshness in the storage or transportation process, the apples usually are coated with waxed on the surface. However, consumers are difficult to distinguish apples whether with the waxed with the naked eyes. At present, the mainstream methods of detecting fruit waxed are gas-mass-linked [1], high-performance liquid chromatography [2, 3], etc. These methods are time-consuming and require high skill for operators.

In recent years, hyperspectral imaging technology has developed rapidly. Zhao et al. used hyperspectral imaging techniques and multivariate correction methods to detect the hardness of apples [4]. Liu et al. used hyperspectral imaging techniques to combine the wavelength selection method with the partial least squares (PLS) method for mutagenic nondestructive testing [5]. Gao et al. used MSC-SPA-LS-SVM model to explore the feasibility of using hyperspectral imaging to identify different apple [6]. However, the scientists have not yet qualified the near infrared hyperspectral images before they were used for classification model training process [7–10]. The low-quality data even the error data always can destroy the whole model training process [11–16].

In order to improve the classification accuracy and stability of the prediction model, we propose a method to evaluate the quality of the hyperspectral image of the sample in this study. The spectral correlation analysis of each pixel was used to determine quality of the sample's hyperspectral image in this study. When the spectral correlation of all the pixels in the sample is too low, the sample data is eliminated before being input to the model training set. The least squares support vector machine (LS-SVM) is an improvement to the traditional support vector machine and is used to establish the classification model between the hyperspectral and the waxed, and the more reliable prediction results can be obtained, which provides a reference for the further study of modeling with the hyperspectral images.

2 Materials and Methods

2.1 Apple Samples Preparation

The samples were selected from 20 apples of "Fushi" variety, which were purchased from ordinary supermarkets on March 24, 2017. Their size and shape were similar and intact. The purchase of 20 apples are waxed object, and randomly selected 10 apples for its hyperspectral image acquisition. And then the other 10 apples were scraped on their surface by knife and the waxed was removed gently without hurting the apples' surface. each scraping one immediately for its hyperspectral image acquisition, collecting each non-waxed apple image position relative to the collection of waxed apples Image position unchanged.

2.2 Hyperspectral Image Acquisition System

As shown in Fig. 1, the hyperspectral image acquisition system used in this experiment consists of the Hyperspec Spectrometer (Headwall Photonics, G4-332), the motor controller box, a precision mobile platform, the illumination light source assembly (including power supply and fiber illuminator-fiber optic quartz tungsten halogen (QTH) lamp). The QTH lamp was fixed at the right upside of the mobile platform at an angle of 45 with the height of 30 cm. The system acquisition software is Headwall Hyperspec III. The detailed information of the hyperspectral imaging system is as following: exposure time: 20 ms, frame rate: 42, image size: 320 * 426, fruit speed: 14.61 mm/s, effective wavelength range: 550–1710 nm.

2.3 Hyperspectral Image Acquisition and Calibration

In order to obtain the high spectral image, the speed of the transport platform and the distance between the lens and samples need to be adjusted. The hyperspectral imaging system worked in reflectance mode and scanned all the samples line by line. All the samples in the experimental set would be imaged with a single acquisition. Thus, a total of twenty hyperspectral images were got for one sample. Each data was calibrated by Eq. (1),

$$R_{data} = \frac{R_{raw} - R_b}{R_w - R_b} \qquad (1)$$

where R_{data} is corrected image, R_w is calibration image obtained by scanning the standard white correction board (reflectance close to 99.9%), R_b is the calibration image when the lens is covered with its cap (reflectance close to 0%). R_{raw} is the original hyperspectral image.

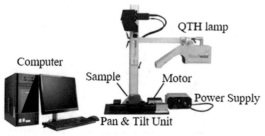

(a) The hyperspectral image acquisition system

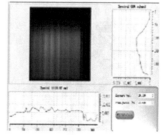

(b) Headwall Hyperspec III

Fig. 1 Schematic diagram of the hyperspectral imaging system

2.4 Data Processing and Analysis

All of the acquired spectral images were processed and analyzed by the Environment for Visualizing Images software program (ENVI 5.1, Research System Inc., Boulder, CO., USA), and the Matlab 2015a (The MathWorks Inc., Natick, USA) image processing toolbox.

The regions of interest (ROI) hyperspectral image data used for modeling training and prediction was chosen from each apple sample hyperspectral image. The size was 28 pixels × 28 pixels. Then all of the ROI was denoised by using the Minimum Noise Fraction Rotation (MNF).

3 Results and Discussion

3.1 Hyperspectral Images of Different Apple Categories

The typical reflectance hyperspectral before and after remove the waxed from the same apple ROI is shown in Fig. 2. Figure 2a is the reflectance hyperspectral image of waxed apple. The red rectangle area is one of the ROI areas in this study. Figure 2b is the average spectral of this ROI region. The red curve is the spectral of waxed apple. The blue curve is the spectral of non-waxed apple. Taking into account the amplitude and signal-to-noise ratio at the beginning and end of the spectrum was too little, from 550 to 1650 nm spectral range of experimental data was selected for this study's analysis.

The Fig. 2 shows that the trend of hyperspectral curves before and after waxed apples ROI is similar and around the 806 nm wavelength reflectivity reached the highest. The average reflectance of the non-waxed region was lower than that of the waxed region in the whole spectral range. In this study, the peaks and valleys of the reflectance hyperspectral, such as 639.90, 806.85, 973.80, 1073.97, 1197.99, 1269.54 and 1441.26 nm, were picked as the characteristic points for further analysis.

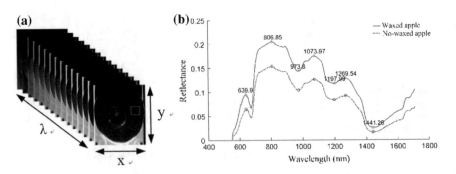

Fig. 2 The typical reflectance hyperspectral from the two waxed apples ROI

3.2 Spectral Correlation Analysis

The detected samples like apple always have not absolutely uniform ROI region. However, the low quality, even the error spectral of sample ROI can destroy the modeling process and get a non-stable predicted model. Thus, the quality of the training and predicted spectral samples is evaluated before data bringing into the model training and predicting process. In this study, the correlation degree of each pixel in the ROI was taking into account as a criterion for evaluating the samples' quality. The spectral of each pixel in the ROI should be uniform with strong correlation.

In the selected ROI region, two effective wavelengths (806.85 and 1073.97 nm) of ROI can be used as the object of study. The pixel correlation is analyzed as shown in Fig. 3.

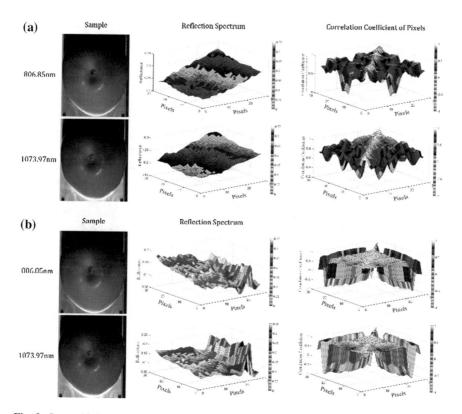

Fig. 3 Spectral information and pixel correlation coefficient of different samples

Figure 3a shows the reflection spectra and pixel correlation coefficients of high quality samples. The reflectance spectra of the high-quality samples are smoother and the pixel correlation coefficients are mostly greater than 0.5. At 806.85 and 1073.97 nm, the ratio of pixels in the samples with a correlation coefficient greater than 0.5 are 0.95 and 0.97. Figure 3b shows the reflection spectrum and the pixel correlation coefficient of the low-quality sample. The reflectance spectra of the low-quality samples are quite different, and even the pixel correlation coefficient is less than 0. At 806.85 nm and 1073.97 nm, the ratio of pixels in the sample with a correlation coefficient greater than 0.5 are 0.73 and 0.70.

Strongly correlated pixels have similar spectral reflectance. According to the above information, the sum of the pixels with the correlation coefficient greater than 0.5 in the sample, which occupies the sample pixels ratio greater than 0.9. This sample is considered a high-quality sample. Then all the samples were collected for pixel correlation calculation, 390 samples of the pixel correlation coefficient greater than 0.9, 10 samples of the pixel correlation coefficient less than 0.9.

3.3 Least Squares Support Vector Machine (LS-SVM) Model

Prediction model of sample spectral data were established by using the least squares support vector machine (LS-SVM). Non-waxed of apples were labeled as 0 (200 samples) and waxed of apples were labeled as 1 (190 samples, 10 low-quality samples). Then, 400 samples were divided into two subsets: one subset containing 300 samples (140 waxed samples, 10 low-quality samples and 150 non-waxed samples) was modeling set. The other subset containing 100 samples (50 waxed samples, 50 non-waxed samples) was used for LS-SVM model on the waxed classification validation modeling. The prediction results of the model were shown in Table 1.

Table 1 Prediction results by LS-SVM model

Number of low-quality samples	Number of high-quality samples	Modeling set	Predictive set
		Correct rate (%)	Correct rate (%)
0	300	99	94
10	290	98	86

4 Conclusion

In our research, waxed and non-waxed of apple as the object of study, the hyperspectral image information of two kinds of apples was obtained by hyperspectral imaging instrument. The original spectral data were processed by matlab software, calculated the correlation between the image pixels. The pixels with the correlation coefficient greater than 0.5 in the sample occupies the sample pixels ratio greater than 90. This sample is considered a high-quality sample. In this way, the quality of the collected samples is screened, and selected 390 high-quality samples and 10 low-quality samples. The prediction accuracy of the two kinds of apples were 94% and 86%, and the low-quality samples affected the prediction accuracy by using LS-SVM model.

The results show that the pixel correlation in this paper by monitoring the quality of the hyperspectral image measured, more reliable prediction results can be obtained, which can make the noninvasive discrimination of food safety come to the practice application sooner.

References

1. Wang, L., Cai, L.S., Lin, Z., Zhong, Q.S., Lv, H.P., Tan, J.F., Guo, L.: Analysis of aroma compounds in white tea using headspace solid-phase micro-extraction and GC-MS. J Tea Sci.
2. Guo, Y., Liang, J., Li, M.M., Zhao, Z.Y.: Determination of organic acids in apple fruits by HPLC. **33**(02), 227—231 (2012)
3. Wenen, Z., Jinqiang, Z., Qiao Xiansheng, Y., Hong, H.: J Fruit Sci. **30**(2), 115–123 (2010)
4. Zhao, J., Liu, J., Chen, Q., et al.: Detecting subtle bruiseson fruit swith hyperspectral imaging. Trans Chin. Soc. Agric. Mach. **39**(1), 106–109 (2008)
5. Liu, Y., Bai, Y., Qiu, Z., Chen, W., Feng, Y.: Hyperspectral imaging technology and wavelength selection method for nondestructive detection of mutton adulteration. J Hainan Norm. Univ. (Natural Science), **28**(3) (2015)
6. Jun-Feng, G., Hai-Liang, Z., Wen-Wen, K., Yong, H.: Nondestructive discrimination of waxed apples based on hyperspectral imaging technology. Spectrosc. Spectr. Anal. **33**(7), 1922–1926 (2013)
7. Lawrence, K.C., Yoon, S.C., Heitschmidt, G.W.: Imaging system with modified-pressure chamber for crack detection in shell eggs. Sens. Instrum. Food Qual. Saf. **2**(3), 116–122 (2008)
8. Baranowski, P., Mazurek, W., Wozniak, J., Majewska, U.: Detection ofearly bruises in apples using hyperspectral data and thermal imaging. J. Food Eng. **110**, 345–355 (2012)
9. Yu, K., Zhao, Y., Liu, Z., Li, X., Liu, F., He, Y.: Application of visible and near-infrared hyperspectral imaging for detection of defective features in Loquat. Food Bioprocess Technol. **7**, 3077–3087 (2014)
10. Kamruzzaman, M., ElMasry, G., Sun, D.-W., Allen, P.: Prediction of some quality attributes of lamb meat using near-infrared hyperspectral imaging and multivariate analysis. Anal. Chim. Acta **714**, 57–67 (2012)
11. Rivera, N.V., Gómez-Sanchis, J., Chanona-Pérez, J., Carrasco, J.J., Millán-Giraldo, M., Lorente, D., Cubero, S., Blasco, J.: Early detection of mechanical damage in mango using NIR hyperspectral images and machine learning. Biosyst. Eng. **122**, 91–98 (2014)
12. Xiaoli, L., Pengcheng, N., Yong, H., et al.: Expert Syst. Appl. **38**(9), 11149 (2011)

13. Lu, H., Li, Y., Chen, M., Kim, H., Serikawa, S.: Brain intelligence: go beyond artificial intelligence. Mob. Netw. Appl. 1–10 (2017)
14. Lu, H., Li, Y., Mu, S., Wang, D., Kim, H., Serikawa, S.: Motor anomaly detection for unmanned aerial vehicles using reinforcement learning. IEEE Internet Things J. https://doi.org/10.1109/JIOT.2017.2737479
15. Lu,H., Li, B., Zhu, J., Li, Y., Li, Y., He, L., Li, J., Serikawa, S.: Wound intensity correction and segmentation with convolutional neural networks. Concurr. Comput. Pract. Exp. https://doi.org/10.1002/cpe.3927
16. Serikawa, S., Lu, H.: Underwater image dehazing using joint trilateral filter. Comput. Electr. Eng. **40**(1), 41–50 (2014)

A Diffeomorphic Demons Approach to Statistical Shape Modeling

Guangxu Li, Jiaqi Wu, Zhitao Xiao, Huimin Lu, Hyoung Seop Kim and Philip O. Ogunbona

Abstract Automatic segmentation of organs from medical images is indispensable for the applications of computer-aided diagnosis (CAD) and computer-assisted surgery (CAS). Statistical Shape Models (SSMs) based scheme have been proved as the accurate and robust methods for extraction of anatomical structures. A key step of this approach is the need to place the sampled points(landmarks) with correspondence across the training set. On the one hand, the correspondence of landmarks is related the quality of SSMs. On the other hand, in many cases the location of key landmarks should be manipulated by physicians, since an unattended system is hard to use in most clinical applications. In this paper, we establish a dense correspondence across the whole training set automatically by surface features, which are registered using diffeomorphic demons approach. And the optimization is executed on spherical domain. We establish the SSM for lung regions, the deformation of where is greatly. Finally, we derive quantitative measures of model quality and comparison of segmentation results using the model with non-optimized correspondence.

G. Li (✉) · J. Wu · Z. Xiao
School of Information and Communication Engineering,
Tianjin Polytechnic University, Tianjin, China
e-mail: liguangxu@tjpu.edu.cn

J. Wu
e-mail: wujiaqi@tjpu.edu.cn

Z. Xiao
e-mail: xiaozhitao@tjpu.edu.cn

H. Lu · H.S. Kim
Department of Control Engineering, Kyushu Institute of Technology, Fukuoka, Japan
e-mail: lu@cntl.kyutech.ac.jp

H.S. Kim
e-mail: kim@cntl.kyutech.ac.jp

P.O. Ogunbona
School of Computing and Information Technology, University of Wollongong,
Wollongong, NSW, Australia
e-mail: philipo@uow.edu.au

© Springer International Publishing AG 2018
H. Lu and X. Xu (eds.), *Artificial Intelligence and Robotics*,
Studies in Computational Intelligence 752,
https://doi.org/10.1007/978-3-319-69877-9_14

Keywords Statistical shape model · Landmarks correspondence · Mesh registration · Computer-assisted surgery

1 Introduction

Statistical shape models (SSMs) are successfully used in the medical image segmentation on account of their high accuracy and robustness to noise. To build such an expressible shape model, it is difficult to get away from the issue of correspondence. Generally, the first stage of correspondence is projecting the training samples onto points into a continuous shape space. The model probability density function (pdf) is then defined on this space. And a rule for assigning a point-to-point correspondence between the shapes can be then defined. A simple approach to generate 3-D correspondence is to equally space a fixed number of points on each surface, effectively establishing a correspondence through shape-preserving parameterization. However, such approaches can lead to poor results. Hence, various optimization strategies have been proposed.

Kelemen et al. describe a method of building shape models from a set of closed 3-D surfaces by defining correspondence through a spherical harmonic parameterization of each shape [8]. They pose surface parameterization as an optimization problem by finding the mapping, from the surface to a sphere. Eck et al. employs a 3D intensity model based on spherical harmonics, which analytically describes the shape and intensities of the foci [3].

A principled approach to establishing correspondence is to treat the task as a groupwise optimization problem. The essential solution to this method is choosing a property optimization to global shape. A significant work is from Davies et al. They pose model building as an explicit optimization task [2]. An objective function based on the MDL principle is explicitly optimized using genetic algorithm search, with respect to a piecewise linear representation of the training. They also improved the representation of using a set of kernel functions. Based on the MDL framework, many novel improvement have been established to make the optimization procedure are more easier and faster [1, 6].

In many clinical applications of CASs [11, 15], which uses SSMs to assist the segmentation of target organs. The correspondence is driving by surface features, which is similar as the works [4, 12]. Generally, the physicians set some key landmarks manually. Then the SSM could be quickly located to confirm the finial surface of target. We propose a pairwise strategy to satisfy simultaneously the model quality as well as the landmarks manipulation. In our previous modeling works [7, 9], the landmarks are generated from a reference sphere whose landmarks are distributed according to the surface curvature. However, since the surfaces of training samples are not always aligned well, in this paper we introduce demons algorithm to register local character details of the model surface.

2 Method

2.1 Manipulating Correspondence in 3-D

As noted above, the utility of SSMs depends on the appropriateness of the set of parameterizations, and the correspondence established between the training set. The flowchart of our method is shown in (Fig. 1). Firstly, one instance of training samples is selected as the reference sample. We remesh it according to the surface curvature, and use the vertices of simplified mesh as the landmark positions. Then, all the instances as well as the simplified mesh are transformed to the spherical parametric domain by Spherical Conformal Mapping method [5]. Thirdly, a spherical registration method based on the diffeomorphic demons is utilized to align the training samples in parameterization domain. The details of registration is shown in the following section. Finally, searching for coordinates of landmarks on the aligned spherical maps in terms of the vertices of spherical map of the simplified mesh. Tracked back to the instance surfaces, the locations of landmarks could be confirmed correspondingly.

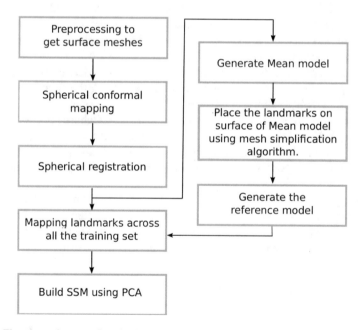

Fig. 1 Flowchart of proposed method

2.2 Principle of Spherical Registration

Let R be the referenced fixed spherical mesh, F be the floating spherical mesh and Γ be the desired transformation that deforms the floating mesh F to match the referenced mesh R. We introduce a hidden transformation c and seek

$$(c^*, s^*) = \underset{c,s}{\text{argmin}} \left\| \sum{}^{-1} (R - F \circ s) \right\|^2 + \delta^2 \text{dist}\,(c, s) + \mu^2 \text{Reg}\,(c) \qquad (1)$$

In this case, the reference mesh R and warped float mesh $F \circ s$ are treated as $N \times 1$ vectors. Typically, $\text{dist}\,(c, s) = \|c - s\|^2$, encouraging the resulting transformation s to be close to the hidden transformation c and $\text{Reg}\,(c) = \|\nabla (c - Id)\|^2$, i.e. the regularization penalizes the gradient magnitude of the displacement field $c - Id$ of the hidden transformation c. δ and μ provide a tradeoff among the different terms of the objective function. Σ is typically a diagonal matrix that models the variability of a feature at a particular voxel. It can be set manually or estimated during the construction of an atlas.

In the classical Diffeormorphic Demons algorithm, the update μ is a diffeormorphism from $\mathbb{R}^3$ to $\mathbb{R}^3$ parameterized by a stationary velocity field v. Note that v is a function that associates a tangent vector with each point in $\mathbb{R}^3$. Under certain mild smoothness conditions, a stationary velocity field v is related a diffeomorphism through the exponential mapping $u = exp\,(v)$.

This formulation facilitates a two-step optimization procedure that alternately optimizes Eq. (1). Starting from an initial displacement field, the Demons algorithm iteratively initial displacement field seeks an update transformation to be composed with the current estimate. In [14], Yeo et al. demonstrate suitable choices of dist (c, s) and Reg (c) that lead to efficient optimization of the modified Demons objective function in Eq. (1). Here, we construct updates u as diffeomorphisms from S^2 to S^2 parameterized by a stationary velocity field $\vec{v}$, which is a tangent vector field on the sphere.

We mapped the lung lobe mesh models into unit spheres. And calculate the mean value of surface curvatures. Comparing the results before/after registration, we can see that curvature textures are more yielding and consistent (Fig. 2), from which we can estimate the mesh surfaces become more uniform after registration.

2.3 Multiscale and Realocate

The algorithm runs in a multi-scale registration. On the low level, we use more "smoothed" surface. These surfaces are generated by an iterative smoothing approximation, and the difference is iterate time. The reason is that the more smoothed surface represents the global shape. Moreover, we sample the vertices on spherical surface using a subdivided icosahedral mesh. In our work, we begin from an icosahedral mesh that contains 2562 vertices and work up to a subdivided icosahedral mesh with the level incremented.

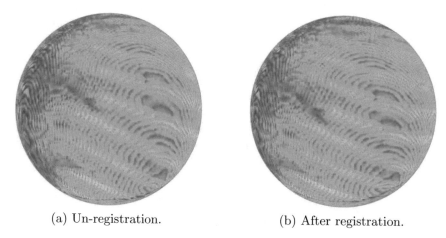

(a) Un-registration.   (b) After registration.

Fig. 2 Mean spherical mapping. The regions in yellow color represent high curvature, while the blue parts represents low curvature regions

3 Experiments

3.1 Data Preprocessing

For our study, 21 cases of thoracic computed tomography (CT) from the LIDC-IDRI (Lung Image Database Consortium and Image Database Resource Initiative) are used. The size of the images varies from to voxels and the gap between the voxels. First, the CT images are pre-processed to enhance the lung region. Second, the region of interest (ROI) is extracted from the CT images by using the region growing method and represented by the binary voxel data. Third, the Marching Cubes algorithm is employed to generate a surface mesh while maintaining the inter-slice connectivity. Finally, a smoothening mean filter is applied to each surface to remove its roughness. In out experiment, the surfaces consist of 69676 vertices and 139349 faces on average. After the correspondence, each mesh obtained by the classic method and our method both contain 642 vertices and 1280 faces.

3.2 Model Evaluation

In order to compare different correspondence methods, in [13] the authors established a consistent and objective basis for comparing models. Here, we use criteria that are based on model properties:

- generalization ability, which measures the ability to represent unseen instances of the class of object modeled.

- specificity, assumes that the SSM should only generate instances of the object class that are similar to those in the training set.

Each measure is evaluated for models containing a varying number of modes of variation, ordered by variance. The standard error for each evaluation can also be calculated, allowing meaningful comparisons between different approaches. So for each dataset, an SSM was constructed using the proposed algorithm, and compared to a model built using non-optimized method of Honda et al. [7]. So, for two of the datasets, we obtain a good approximation to the method by equally spacing points over the training samples. The results are listed below.

3.3 Segmentation Experiments

The segmentation using SSM can be regarded as a matching between the transformed model and the target structure of ROI in the image. This process can be summarized into two steps: In the first step, a shape instance of the model can be expressed by transformation T and parameter b_s. The Mahalanobis distance defined in the appearance model is calculated along the norm vector of each landmark. In each norm, the point with the lowest Mahalanobis distance among the candidate points is selected as the new position of the landmark. The displacement vector of the positions of the landmarks can be expressed as dy_p. In the second step, the generalized match is performed on and to find a new transformation and new residual displacement. These two steps are applied iteratively until a convergence criterion is met, e.g. the variation of movement is under a threshold or the iteration is up to a maximum. The Dice similarity coefficient (DSC) method is used to evaluate the segmentation result with their volumetric overlap. The gold standard data are obtained referencing segmented results from radiologist. The results are listed in (Fig. 4).

3.4 Discussion

The results shown in (Fig. 3) give a quantitative comparison of proposed modeling approach to the previous one; they show that the proposed models have better generalization and specificity properties than the previous model for the datasets. The generate model has a similar compactness for the dataset. A key assumption of the method presented here is that the positions of landmarks could be reconfigurable with a property number of landmarks. The approach that has been taken is based on the intuition that the parts of surface with the similar local geometry should be correspond together.

From the segmentation accuracy results in (Fig. 4), we can see a significantly drop-down on left lobes as well as the right lobes. The reason is that the chest CT data we used is from LIDC-IDRI database, which includes some badly damaged

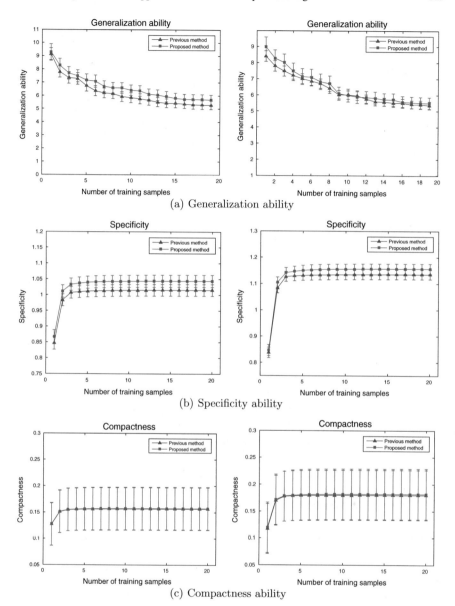

(a) Generalization ability

(b) Specificity ability

(c) Compactness ability

Fig. 3 Quantitative comparison of the lung models. The left column represents left lobe models; The right column is generated by the right lobe models

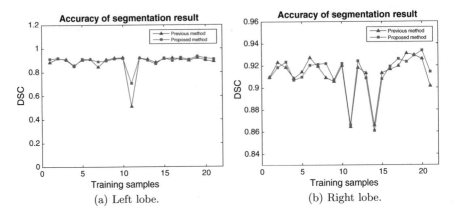

Fig. 4 Segmentation accuracy of lung regions

image regions caused the lesions. Except these cases, the mean value of segmentation accuracy could be promoted obviously.

4 Conclusion

In this paper, we presented a landmark correspondence method for the construction of SSM. We have described a principled and practical approach to automatically construct 3-D statistical shape models. We have shown that the method produces models that have similar compactness, generalization, and specificity properties than previous methods. By referencing the gradient information in the registration process, a compact model is constructed. The experiment result shows that the model constructed by the proposed method and the previous method are almost the same. However, the segmentation accuracy is improved by the proposed method. Future work will consider the use of more instances to train the model and improve its quality. In addition, we will introduce the convolutional neural networks [10] as post-processing of SSM based segmentation for higher accuracy.

Acknowledgements The support for this research is provided in part by the Tianjin Application Foundation and Frontier Technology Research Project (14JCYBJC42300).

References

1. Bernard, F., Vlassis, N., Gemmar, P., Husch, A., Thunberg, J., Goncalves, J., Hertel, F.: Fast correspondences for statistical shape models of brain structures. In: Proceedings of the SPIE, vol. 9784, pp. 97840R–97840R–8 (2016)

2. Davies, R.H., Twining, C.J., Cootes, T.F., Waterton, J.C., Taylor, C.J.: 3D Statistical Shape Models Using Direct Optimisation of Description Length, pp. 3–20. Springer, Heidelberg (2002)
3. Eck, S., Wörz, S., Müller-Ott, K., Hahn, M., Biesdorf, A., Schotta, G., Rippe, K., Rohr, K.: A spherical harmonics intensity model for 3d segmentation and 3d shape analysis of heterochromatin foci. Med. Imag. Anal. **32**, 18–31 (2016)
4. Garcin, L., Rangarajan, A., Younes, L.: Non rigid registration of shapes via diffeomorphic point matching and clustering. In: 2004 International Conference on Image Processing, 2004. ICIP '04, vol. 5, pp. 3299–3302 (2004)
5. Gu, X., Wang, Y., Chan, T.F., Thompson, P.M., Yau, S.: Genus zero surface conformal mapping and its application to brain surface mapping. IEEE Trans. Med. Imag. **23**(8), 949–958 (2004)
6. Heimann, T., Wolf, I., Meinzer, H.P.: Automatic generation of 3d statistical shape models with optimal landmark distributions. Methods Inf. Med. **46**(3), 275–281 (2007)
7. Honda, H., Kim, H., Tan, J.K., Ishikawa, S.: Liver Segmentation for Contrast-enhanced Abdominal mr Images Using Graph Cuts Algorithm (2013)
8. Kelemen, A., Gábor, S., Guido, G.: Elastic model-based segmentation of 3-d neuroradiological data sets. IEEE Trans. Med. Imag. **18**(10), 828–839 (1999)
9. Li, G., Honda, H., Yoshino, Y., Kim, H., Xiao, Z.: A supervised correspondence method for statistical shape model building. In: 2016 IEEE International Conference on Signal and Image Processing (ICSIP), pp. 37–40 (2016)
10. Lu, H., Li, B., Zhu, J., Li, Y., Li, Y., Xu, X., He, L., Li, X., Li, J., Serikawa, S.: Wound intensity correction and segmentation with convolutional neural networks. Concurr. Comput. Pract. Exp. **29**(6), e3927–n/a (2017)
11. Niessen, W.: Model-Based Image Segmentation for Image-Guided Interventions, pp. 219–239. Springer US, Boston, MA (2008)
12. Paulsen, R.R., Baerentzen, J.A., Larsen, R.: Markov random field surface reconstruction. IEEE Trans. Vis. Comput. Graph. **16**(4), 636–646 (2010)
13. Styner, M.A., Rajamani, K.T., Nolte, L.P., Zsemlye, G., Székely, G., Taylor, C.J., Davies, R.H.: Evaluation of 3D Correspondence Methods for Model Building, pp. 63–75. Springer (2003)
14. Yeo, B.T.T., Sabuncu, M.R., Vercauteren, T., Ayache, N., Fischl, B., Golland, P.: Spherical demons: Fast diffeomorphic landmark-free surface registration. IEEE Trans. Med. Imag. **29**(3), 650–668 (2010)
15. Yoshino, Y., Miyajima, T., Lu, H., Tan, J., Kim, H., Murakami, S., Aoki, T., Tachibana, R., Hirano, Y., Kido, S.: Automatic classification of lung nodules on mdct images with the temporal subtraction technique. Int. J. Comput. Assist. Radiol. Surg. (2017)

A Joint Angle-Based Control Scheme for the Lower-Body of Humanoid Robot via Kinect

Guanwen Wang, Jianxin Chen, Xiao Hu and Jiayun Shen

Abstract To control humanoid robot, one of the challenges is to keep the balance. In this paper, a joint angle-based control (JAC) scheme is proposed for the lower-body of humanoid robot NAO to imitate human motion via Kinect. To keep the balance, the joint angles in the lower-body of NAO are optimized according to the tracked human motion and the current state of the robot. Experimental results show that the control scheme works efficiently even when the humanoid robot performs the complex movement such as standing on single foot.

Keywords Humanoid robot · Kinect · Control · Lower-body
Joint angle

1 Introduction

Humanoid robots are popular in recent years as they can be applied in many fields, such as smart housing, education, industry, etc. Especially, the humanoid robot is controlled to imitate human motion in real time. About it, previous studies focused on the robot to perform some predefined actions based on the motion database [1]. To imitate human motion in real time, the MVN system from Xsens was used to track human motion [2]. But this equipment is inconvenient, and rather expensive.

G. Wang (✉) · J. Chen (✉) · X. Hu · J. Shen
Key Lab of Broadband Wireless Communication and Sensor Network Technology,
Nanjing University of Posts and Telecommunications, Ministry of Education, Nanjing, China
e-mail: wgw4524@icloud.com

J. Chen
e-mail: chenjx@njupt.edu.cn

X. Hu
e-mail: hxiao2014@outlook.com

J. Shen
e-mail: nino06177@163.com

© Springer International Publishing AG 2018
H. Lu and X. Xu (eds.), *Artificial Intelligence and Robotics*,
Studies in Computational Intelligence 752,
https://doi.org/10.1007/978-3-319-69877-9_15

Recently, a cheap 3D camera–Kinect sensor makes it possible to track human motion in an easy way [3].

In [4], an interaction system of the humanoid robot NAO was introduced and the upper body of humanoid robot NAO is controlled via Kinect. In [5], Avalos et al. realized the teleoperation of the upper body of NAO and used it as an interactive tool for education. In [6], the head of NAO was controlled to imitate human. In [7], Almetwally et al. focused on the joint angle calculation according to the skeleton information from Kinect and achieved the control accuracy of 84%. In [8], three methods for NAO control are discussed to imitate human upper body motions. These studies focused on the upper-body control of humanoid, but for the lower-body control, little work has been done. For the lower-body control of humanoid robot, how to keep the balance is a big challenge. In [9], the balance of robot is maintained on the slope by establishing the dynamic model and applying the linear quadratic regulator method. In [10], Wang et al. formed the trajectory of each degree of freedom to maintain the balance on the ground. However, the structural limits make the ankle strategy impractical and increase the dynamical distortions [11–14]. In this paper, we propose a joint angel control (JAC) scheme to tackle the problem of balance on the ground.

The rest of this paper is organized as follow. In Sect. 2, we introduce the control system of humanoid robot via Kinect. In Sect. 3, we propose a solution for the lower body control of the robot. Experimental results verify the efficiency of the proposed scheme in Sect. 4. And Sect. 5 draws the conclusion.

2   System Overview

Figure 1 depicts the system of the remote control for humanoid robot via Kinect. It consists of three parts: motion tracking, processing/transmission, and real-time robot control. Human motion is tracked by Kinect sensor. Then the data of tracked motion is transmitted to the robot via network (e.g. TCP/IP) after processing. Finally, the control is performed on the Humanoid robot. In our system, we use the humanoid robot NAO [15]. There are two coordinate systems defined for NAO as in Fig. 2: FRAME ROBOT and FRAME WORLD. In the FRAME ROBOT

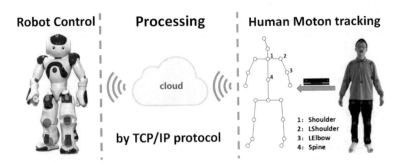

Fig. 1 Description of the control system

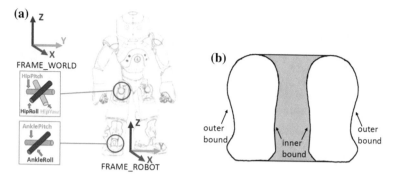

Fig. 2 Two coordinate systems in NAO and the area of support polygon

coordinate, the origin is at the middle of two feet, and X-axis is forwards. In the FRAME WORLD coordinate, the origin is a fixed point that never changes. In the following, we will use these two coordinates to control the lower body of NAO robot.

3 Joint Angle-Based Control Scheme

To let the humanoid robot perform the same action as human, the main task is to control the joints of robot. For lower-body control, the related joints are hip(left/right), knee(left/right) and ankle(left/right). Each of them consists of several angles depending on the degree of freedom (DOF). For example, in NAO there are three joint angles defined for hip, e.g. HipPitch, HipRoll and HipYaw. Two joint angles are defined for ankle: AnklePitch and AnkleRoll (Fig. 2a). Generally, these joint angles can be calculated according to the skeleton data of human motion tracked by Kinect as following

$$\alpha = \arccos\left(\frac{\mathbf{v}_1 \cdot \mathbf{v}_2}{\|\mathbf{v}_1\| \cdot \|\mathbf{v}_2\|}\right). \tag{1}$$

Here v1, v2 are two vectors in space, and α is the joint angle between them. But when we use them directly, some of the angles seem unfeasible to keep the balance of robot, especially for the actions of squatting and standing on one foot.

To combat this problem, we propose a control scheme as in Fig. 3. Firstly, human motion is tracked via Kinect. Then the state of robot is tracked and joint angles are calculated. Finally, the gain factor is defined for the difference between the tracked joint angles of robot and those of the human. This gain factor will be used to optimize the control of robot to keep balance during the movements.

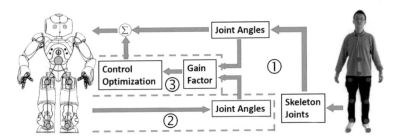

Fig. 3 Description of control concerning the lower-body

3.1 Joint Angles

To keep the balance of NAO, we need to keep the Center of Mass (CoM) of robot inside the support polygon of the robot. The support polygon is determined by the foot when standing. Figure 2b depicts the support polygon when the robot stands with two feet. If the robot stands with one foot, the support polygon is just the foot bottom. Therein, for the lower-body control, our solution tries to keep CoM inside the support polygon during movements.

As we discussed above, if we use the all joint angles computed with Eq. 1 to control the robot, it is difficult to keep the balance. Then we have to find a new solution to compute these joint angles with the information from the robot. In the following, we give an example to compute HipRoll in robot, which will be used for the action of standing on one foot.

Figure 4 depicts the scenario when NAO stands on one foot. Theoretically, it will not fall down as long as the CoM is within the area of support polygon. Corresponding to the support polygon, there are two movement bounds: the inner bound and the outer bound. The former is the inner side of the support polygon, and the latter is the outer side of the support polygon. Therein, to keep the balance of robot, the CoM should be kept inside between these two bounds. Related to these two bounds, there are two angles β_L and β_H for HipRoll as in Fig. 4b and c. Herein, to let NAO stand on one foot is to control the joint angle HipRoll satisfying $\beta_L \leq \beta_{HipRoll} \leq \beta_H$. According to [10], when the projection of CoM falls on the line connecting the left and right ankle joints, the robot will be in the stable condition. Thus during the standing on one foot, when CoM projects on the ankle joint, it is the optimal state as in Fig. 4a. In this scenario, the joint anglse HipRoll and AnkleRoll are equal (In Fig. 4a, $\beta_c = \theta_{AR}$. β_c is HipRoll and θ_{AR} is AnkleRoll).

For example, we want NAO to stand on the right leg. We use β to represent the joint angle RHipRoll. According to the physical limitation of NAO, we confine the range of RHipRoll to $0° < \beta < 21.7°$. As the distance between HipRoll and AnkleRoll is the length of leg, we assume it L_{RLeg}. Also, L_{CH}^R denotes the distance between CoM and RHipRoll. When the projection of CoM falls within the area of

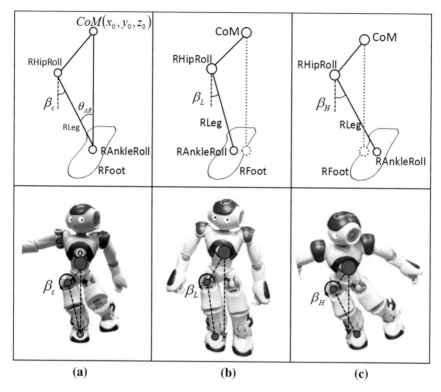

Fig. 4 The model of Roll in single-foot mode

RAnkelRoll, it will be optimal for NAO standing on the single foot. In this scenario, the value of RHipRoll is the optimal angle β_c, which is

$$\theta_{RHipRoll} = \beta_c = \arccos\left(\frac{z_0^2 + L_{RLeg}^2 - L_{CH}^{R\ 2}}{2z_0 \cdot L_{RLeg}}\right) \qquad (2)$$

Similarly, β_L and β_H can also be obtained using this method.

3.2　Gain Factor

After we obtain the joint angles of robot, we compare them with those from human. Let CoM move horizontally, we assume two angle sequences $\Theta_H = \{\theta_{H1}, \theta_{H2}, \theta_{H3}, \ldots, \theta_{HN}\}$ and $\Theta_R = \{\theta_{R1}, \theta_{R2}, \theta_{R3}, \ldots, \theta_{RM}\}$ for the joint angle hip on human and robot respectively. N and M denote the length. Then the correlation coefficient of these two sequences is

$$\rho_{H,R} = \frac{\sum_{i=0}^{n} \left(\theta_{H,i} - \overline{\theta_H}\right)\left(\theta_{R,i} - \overline{\theta_R}\right)}{\sqrt{\sum_{i=0}^{n} \left(\theta_{H,i} - \overline{\theta_H}\right)^2}\sqrt{\sum_{i=0}^{n} \left(\theta_{R,i} - \overline{\theta_R}\right)^2}} \qquad (3)$$

where n = min{M, N}. $\rho_{H,R} \in [-1,1]$. If there is $\rho_{H,R} > 0$, these two sequences are in the positive correlation. We define a gain factor related to the mapping from human to robot as

$$A_\omega = \frac{\sum_i \frac{\theta_{i,R}}{\theta_{i,H}}}{\min\{M,N\}}.$$

This gain factor will be use to optimize the control of robot to keep the balance.

3.3   Control Optimization

With this gain factor, the mapping from human to robot can be optimized. However, when we compensate with this gain factor, the joint angles may change quickly, which might cause the robot to lose balance due to the inertia during the movement. To combat it, we design an optimized method to control the changing rate of joint angles. In this method, we use a multistep incremental curve to replace the original extracted data. Each time when NAO receives a frame data extracted from the Kinect, we let NAO compare it with the previous one. If the new frame data is greater than the old one, there will be an upright step increase. If it is smaller than the previous one, there will be the opposite decrease step. The initial data is set as the same value of the original one. Then the modified data looks like a continuous staircase going up and down. However, if we minimize the step, the receiving curve may keep smooth. For NAO, there will be no distortion if the bit rate is high enough.

When NAO moves quickly, the slope overload will occur, which is similar to the condition of slope overload in Delta Modulation(ΔM). Slope overload means that the new modified data curve cannot catch up with the original one, which in turn realizes the goal of slowing down the changing rate of data. And this helps NAO to keep the balance and move stably.

4   Experiment Evaluation

In our experiment, human motion is tracked via Kinect V2.0 and NAO is controlled via WLAN in the lab. The computer has an Intel i7 CPU of 2.60 GHz and the GTX960m graphics card. Figure 5 depicts the process of standing on one foot.

Fig. 5 Joints received and executed by NAO

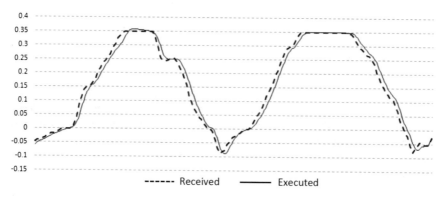

------ Received   ——— Executed

Fig. 6 Joints received and executed by NAO

(1) Angle Matching

In JAC method that we have proposed, we modify the data with gain factor and slow down the changing rate of joint angles. After being sent to NAO, the data of joint angles also needs to be confined within the range of DOFs. Here the matching degree is used to reflect the control accuracy. It is the difference between the received joint angles and the actual values executed by NAO. Figure 6 depicts two sets of joint angle HipRoll. The dashed line represents the data that NAO received, while the solid line represents the data of the joint angles executed by NAO during the experiment(Unit: rad). According to the data shown in Fig. 6, it can be found that two sets of data are close to each other. It means that NAO receives the data and uses the data to perform the action with little distortion. In order to get the matching degree of these two sets of data, we make a lengthways comparison between two sets of data in Fig. 6.

Table 1 Delay analysis

	Angle transmission [16]	Behavioral synchronization [17]	Recognition delay of sEMG-based [18]	Our JAC method
Delay (ms)	1290	400	300	230

Then we can get the deviation between two sets of data: 0.035476, which indicates that the matching degree of the data is 96.45%.

(2) Time Delay

Meanwhile, we make a lateral comparison between these two sets of data in Fig. 6 to get the internal time delay of NAO (Table 1).

From it, we find that our method performs with less time delay than other methods.

5 Conclusion

In this paper a joint angle-based control (JAC) scheme is proposed for the control of the humanoid robot NAO. A new calculation method of joint angles on the lower body of NAO is designed. We introduce a gain factor to help the mapping from human to NAO during the control process. Then we modify the data with control optimization, which slows down the changing rate of joint angles and ensures the stability during the control process.

Acknowledgements This work was supported by the Technique Innovation Training Program (No. 201610293001Z), the open research fund of Key Lab of Broadband Wireless Communication and Sensor Network Technology (Nanjing University of Posts and Telecommunications, Ministry of Education, JZNY201704), funding from Nanjing University of Posts and Telecommunications (NY217021), National Natural Science Foundation of China (Grant No. 61401228), China Postdoctoral Science. Foundation (Grant No. 2015M581841), and Postdoctoral Science Foundation of Jiangsu Province (Grant No. 1501019A).

References

1. Indrajit, W., Muis, A.: Development of whole body motion imitation in humanoid robot. In: International Conference on QiR, pp. 138–141. IEEE (2013)
2. Koenemann, J., Burget, F., Bennewitz, M.: Real-time imitation of human whole-body motions by humanoids. In: International Conference on Robotics & Automation(ICRA), pp. 2806–2812. IEEE (2014)
3. Microsoft Kinect V2.0. Avaliable: www.kinectforwindows.com

4. Csapo, A., Gilmartin, E., Grizou, J., et al.: Speech, gaze and gesturing: multimodal conversational interaction with NAO robot. In: 3rd International Conference on Cognitive Infocommunications, pp. 667–672. IEEE (2012)
5. Avalos, J., Cortez, S., Vasquez, K., Murrary, V., Ramos, O.: Teleoperation using the kinect sensor and NAO robot. In: VII Latin American Symposium on Circuits and Systems (LASCAS), pp. 303–306. IEEE (2016)
6. Li, C., Yang, C.G., Liang, P.D., Cangelosi, A., Wan, J:. Development of kinect based teleoperation of NAO robot. In: 2016 International Conference on Advanced Robotics and Mechatronics (ICARM), pp. 133–138. IEEE (2016)
7. Almetwally, I., Mallem, M.: Real-time tele-operation and tele-walking of humanoid robot NAO using kinect depth camera. In: 10th International Conference on Networking, Sensing and Control(ICNSC), pp. 463–466. IEEE (2013)
8. Mukherjee, S., Paramkusam, D., Dwivedy, S.K.: Inverse kinematics of a NAO humanoid robot using kinect to track and imitate human motion. In: International Conference on Robotics, Automation, Control and Embedded Systems C RACE 2015, pp. 1–7. IEEE (2015)
9. Liang, D.K., Sun, N., Wu, Y.M., Fang, Y.C.: Modeling and motion control of self-balance robots on the slope. In: 31st Youth Academic Annual Conference of Chinese Association of Automation, pp. 93–98. IEEE (2016)
10. Wang, F., Tang, C., Ou, Y.S., Xu, Y.S.: In: Proceedings of the 10th World Congress on Intelligent Control and Automation, pp. 3692–3697. IEEE (2012)
11. Lu, H., Li, Y., Chen, M., Kim, H., Serikawa, S.: Brain intelligence: go beyond artificial intelligence. Mob. Netw. Appl., pp. 1–10 (2017)
12. Lu, H., Li, Y., Mu, S., Wang, D., Kim, H., Serikawa, S.: Motor anomaly detection for unmanned aerial vehicles using reinforcement learning. IEEE Internet Things J. https://doi.org/10.1109/JIOT.2017.2737479
13. Lu, H., Li, B., Zhu, J., Li, Y., Li, Y., He, L., Li, J., Serikawa, S.: Wound intensity correction and segmentation with convolutional neural networks. Concurr. Comput. Pract. Exp. https://doi.org/10.1002/cpe.3927
14. Serikawa, S., Lu, H.: Underwater image dehazing using joint trilateral filter. Comput. Electr. Eng. **40**(1), 41–50 (2014)
15. Aldebaran Robotics NAO. Avaliable: www.ald.softbankrobotics.com
16. Afthoni, R., Rizal, A., Susanto, E.: Proportional derivative control based robot arm system using microsoft kinect. In: International Conference on Robotics and Intelligent Computation Systems (ROBIONETICS), pp. 24–29. IEEE (2013)
17. Igorevich, R., Ismoilovich, E., Min, D.: Behavioral synchronization of human and humanoid robot. In: 8th International Conference on Ubiquitous Robot and Ambient Intelligence (URAI), pp. 655–660. IEEE (2011)
18. Wang, B.C., Yang, C.G., Xie, Q.: Human-machine interfaces based on EMG and kinect applied to teleoperation of a mobile humanoid robot. In: 10th World Congress on Intelligent Control and Automation, pp. 3903–3908. IEEE (2012)

Research on Indoor Positioning Technology Based on MEMS IMU

Zhang Tao, Weng Chengcheng and Yan Jie

Abstract A shoe-mounted indoor positioning system is designed based on the low-cost MEMS (Micro-Electro-Mechanical System) inertial measurement unit (IMU). To solve the problem of the high noise and drift of the MEMS inertial sensor, the method of the coarse calibration for MIMU is studied. Aiming at solving the shortcomings of the traditional zero velocity interval detection algorithm with single threshold, an alternative algorithm for the zero velocity interval detection based on multiple gaits is designed, which judges the motion state before detection. Then, the Kalman filter algorithm based on the zero velocity update (ZUPT) and zero angular rate update (ZARU) is designed to estimate and compensate the cumulative error of the sensors during walking. Finally, a field test based on a low-cost MEMS IMU is carried out with four different gaits, slow walking, up and down staircase, striding forward and long-distance walking with variable speed. The results show that the positioning error of the proposed method is only 3% of the walking distance.

Keywords MEMS IMU · Zero velocity interval detection · Kalman filtering ZUPT · ZARU

Z. Tao (✉) · W. Chengcheng
School of Instrument Science and Engineering,
Southeast University,
Nanjing, China
e-mail: ztandyy@163.com

W. Chengcheng
e-mail: 284873154@qq.com

Y. Jie
Department of General Research, Beijing Institute
of Electronic System Engineering,
Beijing, China
e-mail: yanjiebj@sina.com

© Springer International Publishing AG 2018
H. Lu and X. Xu (eds.), *Artificial Intelligence and Robotics*,
Studies in Computational Intelligence 752,
https://doi.org/10.1007/978-3-319-69877-9_16

1   Introduction

It is researched that people spend more than 80% of the time one day in the indoor environment. With the development of science and technology, the demand for the location services is growing and the location services are gradually used for the emergency rescue, tracking, monitoring, information push and other scenes [1], especially in large shopping malls, airports, railway stations and other public places with a large stream of people. Accurate indoor location services have become important to the development of LBS technology, therefore, the study of high-precision indoor positioning method is a very valuable research filed.

In the outdoor environment, the GPS can provide accurate location and speed information in real time [2]. However, in the indoor environment such as the city centers, basements and buildings, GPS signal is easy to be obstructed by walls or buildings and multi-path effects may exist indoors [3], so that other alternative methods for indoor positioning should be studied [4–6].

Common positioning solutions include wireless location technology (such as radio frequency identification (RFID)), ultrasonic positioning, Bluetooth positioning, wireless local area network (WLAN), ultra-wideband positioning, etc. These technologies need to install infrastructures in buildings in advance, which have a high cost and cannot meet the requirements of the unknown environments. Therefore, in recent years, many scholars will focus on the autonomous indoor navigation system. MEMS inertial equipment, due to its low cost, small size, strong autonomy and good environmental adaptability, has become the best choice for indoor positioning [7], which is especially suitable for unknown environments.

Usually, IMU-based indoor positioning technology cannot provide accurate position for a long time, because of the accumulation of the IMU errors. An indoor positioning algorithm only based on low-cost MEMS IMU is proposed to improve the accuracy of positioning by using gait detection, zero velocity update (ZUPT) and zero angular rate update (ZARU).

2   Indoor Positioning Algorithm

2.1   System Structure

Effective correction algorithm is necessary for indoor positioning using MEMS IMU. In order to complete error correction, MIMU should be put on the right place [8]. It is found that there is a zero velocity interval during the walking process. During the interval, the speed of the MIMU is zero theoretically and the velocity error can be corrected. In the experiment, the MIMU is mounted on the instep [9, 10], which is shown in Fig. 1.

In this paper, the inertial navigation method is used to calculate the location information of the pedestrian. Aiming at the problem of error divergence over time,

MIMU

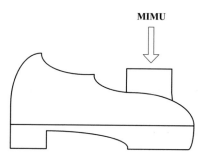

Fig. 1 MIMU mounting method

a precise zero—speed interval detection algorithm adapted to a variety of foot motion states is studied, and Kalman filter is used in the zero-speed interval to correct the error.

The complete indoor positioning process based on MIMU is divided into four steps.

(1) Obtain the initial pitch angle and roll angle information of MIMU, and complete the initial calibration with heading angle provided by the magnetometer.
(2) Calculate the speed and the position of foot with the output data of the MIMU by the conventional strapdown algorithm.
(3) Detect zero velocity interval during walking process by measuring the output data of the MIMU and prepare for error correction algorithm.
(4) The angular velocity of the gyro and the velocity values in the zero velocity interval are used as observed quantity to design extended Kalman filter, then use the result to correct the error.

The specific flow chart is shown in Fig. 2:

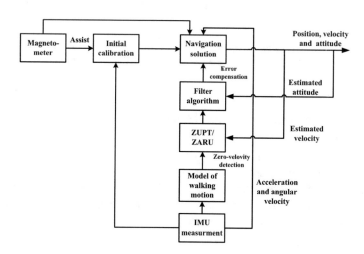

Fig. 2 Flow of the indoor positioning algorithm

2.2 *Initial Calibration*

Low-cost MIMU cannot be regarded as a high accuracy inertial navigation systems. Rational calibration algorithm should be designed to improve the accuracy of the indoor positioning. The pitch angle and roll angle can be expressed as follows [11]:

$$\theta = -\arcsin\left(\frac{f_x^b}{g}\right), \quad \gamma = \arctan\left(\frac{f_y^b}{f_z^b}\right) \tag{1}$$

where f_x^b, f_y^b, f_z^b are components of accelerometer in the x, y, z directions of the body coordination system, and g represents the local gravitational acceleration.

The heading angle can be obtained by the magnetometer output. The outputs of magnetometer are the projection components of the magnetic intensity in the body coordinate system:

$$M^b = \begin{bmatrix} M_x^b & M_y^b & M_z^b \end{bmatrix} \tag{2}$$

The projection components of the magnetic intensity in the navigation coordinate system can be expressed as follows:

$$M_x^n = M_x^b \cos\theta + M_y^b \sin\theta \sin\gamma + M_z^b \sin\theta \cos\gamma \tag{3}$$

$$M_y^n = M_y^b \cos\gamma - M_z^b \sin\gamma \tag{4}$$

The heading angle can be expressed as follows:

$$\varphi = \arctan\left(\frac{M_y^n}{M_x^n}\right) \tag{5}$$

Because the north poles in geographic and the geomagnetic field are not coincident, there is an angle between them, which is called the magnetic declination.

2.3 *Extended Kalman Filter Algorithm*

The errors of the inertial navigation system mainly includes the installation error, biases of accelerometers, drifts of gyros, dynamic inferences, and so on, which lead to the accumulation error, attitude determination error and positioning error [12].

The feet are static during the zero speed intervals, in which the error equations of attitude, velocity and position can be simplified as follows [13]:

$$\begin{cases} \delta\dot{\varphi} = -C_b^n \delta w_{ib}^b \\ \delta\dot{v} = f^n \times \varphi + C_b^n \delta f^b \\ \delta\dot{p} = \delta v \end{cases} \tag{6}$$

where, $\delta\varphi$, δp, δv^n, δf are the attitude errors, position errors, velocity errors and specific force errors respectively, δw_{ib}^b are the gyro angular velocity errors, C_b^n is the attitude matrix.

The foot-mounted positioning system based on MIMU is a nonlinear system, so EKF filter algorithm is used. The state vector of the EKF is shown as follows,

$$\delta x = \begin{bmatrix} \delta\varphi & \varepsilon^b & \delta p^n & \delta v^n & \nabla^b \end{bmatrix}^T \tag{7}$$

where, $\delta\varphi$ are the attitude errors, including the pitch angle error, roll angle error and heading angle error; ε^b are drifts of three gyros; δp^n are position errors in the navigation coordinate system; δv^n are speed errors in the navigation coordinate system; ∇^b are biases of the three accelerometers; According to the state equations of the system, $\delta x_k = \Phi_k \delta x_{k-1} + w_k$, the state transition matrix can be obtained.

$$\Phi_k = \begin{bmatrix} I_{3\times3} & -\Delta t \cdot C_b^n & 0 & 0 & 0 \\ 0 & I_{3\times3} & 0 & 0 & 0 \\ 0 & 0 & I_{3\times3} & \Delta t \cdot I_{3\times3} & 0 \\ \Delta t \cdot f^n & 0 & 0 & I_{3\times3} & \Delta t \cdot C_b^n \\ 0 & 0 & 0 & 0 & I_{3\times3} \end{bmatrix} \tag{8}$$

Theoretically, the speed and the angular velocity are zero in zero-velocity intervals, but actually they are not zero because of the errors of MIMU. In this case, the differences can be regarded as observation variables.

In order to improve the accuracy of the positioning, both ZUPT and ZARU are used and take the velocity and the angular velocity in the zero velocity interval as the observation vector. So the observation vector can be expressed as:

$$Z_k = \begin{bmatrix} c^b & \delta v^n \end{bmatrix}^T = \begin{bmatrix} w^b & v^n \end{bmatrix} \tag{9}$$

The observation matrix can be expressed as:

$$H_k = \begin{bmatrix} 0_{3\times3} & I_{3\times3} & 0_{3\times3} & 0_{3\times3} & 0_{3\times3} \\ 0_{3\times3} & 0_{3\times3} & 0_{3\times3} & I_{3\times3} & 0_{3\times3} \end{bmatrix} \tag{10}$$

Since the observation vectors can only be acquired at zero velocity intervals, time update and measurement update are operated at zero-velocity intervals, while only time update is operated during the remaining gaits.

3 Zero Velocity Detection Algorithm

When a person is walking, gaits can be described as, Heel strike → Foot
flat → Mid-stance → Push off → Toe off [14, 15]. There is a temporarily stationary state when the pace fully contacts with the ground, which called zero
velocity interval. The gaits modes are shown in Fig. 3.

Detecting zero velocity intervals effectively and accurately is significant to
improve the positioning accuracy. Theoretically, during zero-speed intervals, the
modulus of specific forces the gravitational acceleration g. In many cases, it is
unreliable to detect zero velocity intervals only using the acceleration information,
because the foot may be in a uniform linear motion state. Therefore, it is necessary
to introduce other information to assist the judgment.

Indoor walking of a person may have multiple states, such as slow walking,
walking up and down stairs, jumping forward, etc. Different motion states correspond to different outputs of IMU, so that the walking state can be detected by the
measurements of MIMU.

The process of zero velocity detection is as follows.

3.1 Determining the Walking State

The acceleration of MIMU is detected during every walking cycle. By analyzing
the acceleration values, the walking state can be determined by the difference
between the maximum and minimum accelerations. The method of judgement is as
follows.

When $(f_{max} - f_{min} < 3g)$, the walking state is regarded as normal walking.

When $(3g < f_{max} - f_{min} < 4g)$, the walking state is regarded as fast walking.

When $(4g < f_{max} - f_{min} < 7g)$, the walking state is regarded as walking up and
down the stairs.

When $(f_{max} - f_{min} > 7g)$, the walking state is regarded as striding or jumping.

3.2 Zero Velocity Detection

In the conventional detection algorithm, only the acceleration information is used
for detecting the zero velocity interval. According to the result of the judgement, the

Fig. 3 Mode of gaits

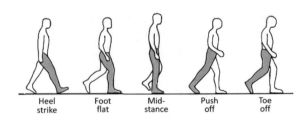

| Heel | Foot | Mid- | Push | Toe |
| strike | flat | stance | off | off |

walking state is complex and the acceleration varies greatly during walking. The conventional detection algorithm with one threshold is not suitable for the complex walking states and may lead to high missed detection rate. To reduce the missed detection rate, various conditions detection algorithms are used in the system, which include the modulus and variance of specific force, the modulus of angular velocity.

- Condition one, the modulus of specific force $|f_k| = \sqrt{f_x^2 + f_y^2 + f_z^2}$. An appropriate confidence interval is set, assuming th_{amin}, th_{amax} are the minimum and maximum values of the confidence intervals respectively,

$$condition\ 1 = \begin{cases} 1 & th_{amin} < |f_k| < th_{amax} \\ 0 & else \end{cases} \tag{11}$$

where 1 represents the stationary state, 0 represents the motion state.
- Condition two, the variance of the specific force. The variance changes little in the zero velocity interval, so the zero velocity interval can be judged by comparing the variance of the specific force to the threshold. The formula is as follows:

$$\sigma_{f_k}^2 = \frac{1}{2m+1} \sum_{j=k-m}^{k+m} (f_j - \bar{f}_k)^2 \tag{12}$$

where, $\bar{f}_k$ is the mean of the specific force; m is the sliding window size.

$$condition\ 2 = \begin{cases} 1 & \sigma_{f_k} < th_\sigma \\ 0 & else \end{cases} \tag{13}$$

- Condition three, the modulus of angular velocity $|w_k| = \sqrt{w_x^2 + w_y^2 + w_z^2}$. During the zero velocity interval, the foot is kept still on the ground and the angular velocity tends to be zero, so the zero-velocity interval is determined by comparing the angular velocity with the threshold th_w.

$$condition\ 3 = \begin{cases} 1 & |w_k| < th_w \\ 0 & else \end{cases} \tag{14}$$

If values of three conditions are one at the same time, the walking state is thought to be in the zero velocity interval.

Figures 4 and 5 show the comparison of conventional detection method with single threshold and improved detection method. It is shown that the conventional detection method cannot detect the zero velocity interval accurately, that could reduce the accuracy of indoor positioning. The improved method greatly improves the accuracy of the detection, which effectively prevents the missed detection and reduces the cumulative error in the process of walking.

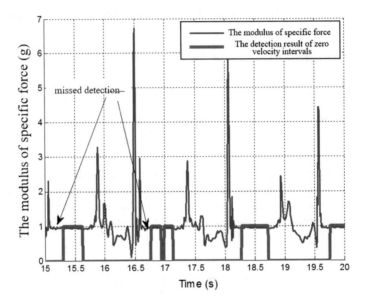

Fig. 4 Test results of the conventional detection method

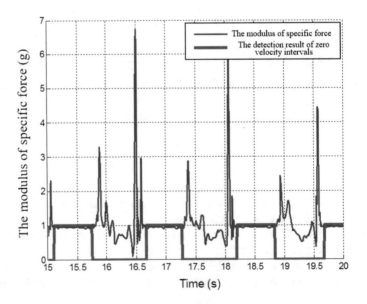

Fig. 5 Test results of the improved detection method

4 Experimental Verification and Result Analysis

4.1 Hardware Introduction

A low cost MIMU is used for experiments, as shown in Fig. 6, which consists of two main parts, an inertial measurement unit (three-axis accelerometers, three-axis gyroscopes) and three-axis magnetometers. The MIMU can connect to a smartphone via Bluetooth. According to the performance of MIMU, the stability of accelerometers and gyros are 0.01 g and 0.05°/s, respectively.

4.2 Results and Discussion

To verify the validity of the proposed algorithm, four kinds of experiments with different walking states are carried out which include normal walking, upstairs and downstairs, walking strides and long distance walking with variable speed. In view of the lack of real position information in the indoor environment, closed loop routes are selected and the distance between starting and the ending position is used to evaluate the positioning accuracy of the experiments.

In order to compare the effect of the correction algorithm, three position curves were drawn for the first experiment. Figure 7 is a position curve without using correction algorithm. Figure 8 is a position curve with conventional single threshold method and ZUPT correction algorithm. Figure 9 is the result of the proposed algorithm. Figures 10, 11 and 12 is the position curves obtained from the experiment 2 to experiment 4.

It can be seen from Figs. 7, 8 and 9, the positioning result without correction algorithm diverges rapidly. The method of conventional single threshold and ZUPT algorithm corrects the error effectively compared to the above method.

Fig. 6 MEMS inertial navigation module

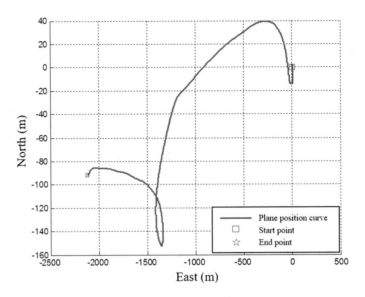

Fig. 7 Result without position correction of experiment 1

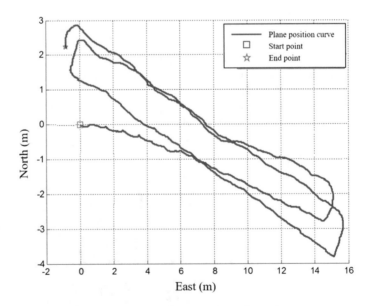

Fig. 8 Result with conventional detection algorithm of experiment 1

However, obvious errors exist when turning around. The proposed algorithm based on ZUPT/ZARU can be adapted to multiple gaits and further reduce the accumulation errors.

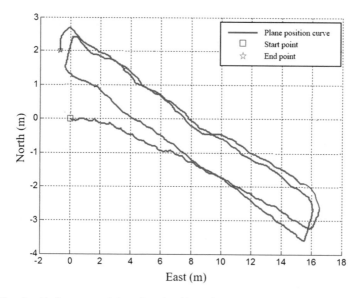

Fig. 9 Result with the proposed detection algorithm of experiment 1

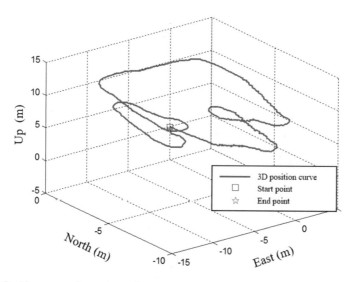

Fig. 10 Position curve of experiment 2

If there are no errors of MIMU, the starting and ending points of the closed loop route in four experiments are coincident. So the distance between the starting and ending points can be regarded as the result of the accumulated errors, when those two points do not coincide. Positioning errors of proposed algorithm are shown as Table 1.

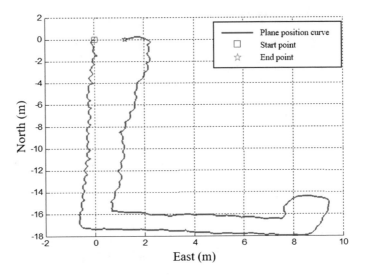

Fig. 11 Position curve of experiment 3

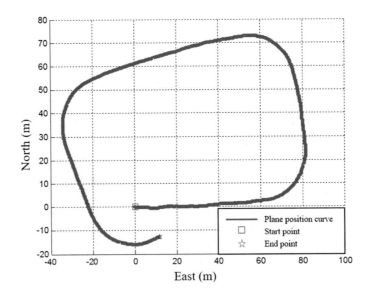

Fig. 12 Position curve of experiment 4

It can be seen that the positioning error of the proposed method is only about 3% of the distance travelled.

Table 1 Positioning error of proposed algorithm

	Walking distance (m)	Offset distance (m)	Percent error (%)
Plane slow walking	73.33	2.16	2.95
Upstairs and downstairs	94.5	2.93	3.10
Stride forward	60.16	1.94	3.23
Long distance walking	340.95	11.10	3.26

5 Conclusion

An indoor positioning method based on MEMS IMU is studied in this paper. The MIMU module is tied on the foot and a zero velocity detection algorithm is designed to adapt to multiple gaits. Then the EKF is used for error correction in the zero velocity intervals. Finally, experiments are carried out, which include four different walking states, normal walking, upstairs and downstairs, stride forward and long distance walking. Results show that the positioning accuracies of the four experiments can reach about 3% of the distance travelled. In the future study, the MIMU system can also be integrated with the indoor map information, such as using wall information to further improve the positioning accuracy.

Acknowledgements This study was supported in part by the National Natural Science Foundation of China (Grant no. 51375088), Foundation of Key Laboratory of Micro-Inertial Instrument and Advanced Navigation Technology of Ministry of Education of China (201403), Fundamental Research Funds for the Central Universities (2242015R30031), Key Laboratory fund of Ministry of public security based on large data structure (2015DSJSYS002).

References

1. Vossiek, M., Wiebking, L., Gulden, P., et al.: Wireless local positioning-concepts, solutions, applications. Radio Wirel. Conf. (RAWCON '03) 219–224 (2003)
2. Jo, K., Chu, K., Sunwoo, M.: Interacting multiple model filter-based sensor fusion of GPS with in-vehicle sensors for real time vehicle positioning. Trans. Intell. Transp. Syst. IEEE **13** (1), 329–343 (2012)
3. Ladd, A.M., Bekris, K.E., Rudys, A.P., et al.: On the feasibility of using wireless ethernet for indoor localization. IEEE Trans. Robot. Autom. **20**(3), 555–559 (2004)
4. Lu, H., Li, Y., Chen, M., Kim, H., Serikawa, S.: Brain intelligence: go beyond artificial intelligence. Mob. Netw. Appl. 1–10 (2017)
5. Lu, H., Li, Y., Mu, S., Wang, D., Kim, H., Serikawa, S.: Motor anomaly detection for unmanned aerial vehicles using reinforcement learning. IEEE Internet Things J. https://doi.org/10.1109/JIOT.2017.2737479
6. Serikawa, S., Lu, H.: Underwater image dehazing using joint trilateral filter. Comput. Electr. Eng. **40**(1), 41–50 (2014)
7. Chun-long, C.A.I., Yi, L.I.U., Yi-wei, L.I.U.: Status quo and trend of inertial integrated navigation system based on MEMS. J. Chin. Iner. Technol. **17**(5), 562–567 (2009)

8. Yun, X., Calusdian, J., Bachmann, E.R., et al.: Estimation of human foot motion during normal walking using inertial and magnetic sensor measurements. Trans. Instrum. Meas. IEEE **61**(7), 2059–2072 (2012)
9. Bebek, O., Suster, M.A., Rajgopal, S., et al.: Personal navigation via shoe mounted inertial measurement units. In: Proceedings of the International Conference on Intelligent Robots and Systems (IROS), Taipei, Taiwan, China, 18–22 Oct 2010
10. Skog, I., Handel, P., Nilsson, J.O., et al.: Zero—Velocity Detection—An algorithm evaluation. Trans. Biomed. Eng. IEEE **57**(11), 2657–2666 (2010)
11. Qi-lian, B.A.O., Zhou-jie, W.U.: Initial alignment methods of low-cost SINS assisted by magnetic compass. J. Chin. Iner. Technol. **16**(5), 513–517 (2008)
12. Choukroun, D., Bar-Itzhack, I.Y., Oshman, Y.: Novel quaternion Kalman filter. IEEE Trans. Aerosp. Electron. Syst. **42**(1), 174–190 (2006)
13. Jin-liang, Zhang, Yong-yuan, Qin, Chun-bo, Mei: Shoe mounted personal navigation system based on MEMS inertial technology. J. Chin. Iner. Technol. **9**(3), 253–256 (2011)
14. Davidson, P., Takala, J.: Algorithm for pedestrian navigation combining IMU measurements and gait models. Gyroscopy Navig. **4**(2), 79–84 (2013)
15. Leppakoski, H., Kappi, J., Syrjarinne, J., et al.: Error analysis of step length estimation in pedestrian dead reckoning. In: Proceedings of the ION GPS, Portland, OR, USA (2002)

Research and Application of WIFI Location Algorithm in Intelligent Sports Venues

Zhang-Zhi Zhao, Meng-Hao Miao, Xiao-Jie Qu and Xiang-Yu Li

Abstract With the development of economic society, comprehensive fitness has risen to national strategy, and people's demand for physical training is getting bigger and bigger. In the information age, traditional stadiums can not meet the needs of people for sports, so the intelligent stadium is the inevitable trend of future development. Also, the positioning of the human in the stadium is the basis for the realization of human-computer interaction and system function. This paper first analyzes the positioning requirements in the intelligent stadium, then puts forward the necessity of the positioning system in the intelligent stadium. The WIFI location fingerprint localization algorithm is further studied and improved while the application of the algorithm in the positioning system of the intelligent stadium is explained. Finally, we completed data collection and real-time positioning in the stadium, designed and implemented the LBS location service based on hospital.

Keywords Intelligent stadium · WIFI positioning system · Location fingerprint LBS

1 Introduction

In October 2014, the State Council issued the document 46 that "Several Opinions of the State Council about Accelerating the Development of Sports Industry to Promote Sports Consumption". The paper puts forward that National fitness rose to national strategy, and by 2025 the total size of sports industry exceeded 5,000 billion Yuan, which is becoming an important force to promote the sustainable development of economic and social development. By then the per capital sports venues will reach 2 square meters, and the number of regular physical exercise will reach 500 million, and sports public services will basically cover the whole people."

Z.-Z. Zhao (✉) · M.-H. Miao · X.-J. Qu · X.-Y. Li
Department of Sports, University of Electronic Science and Technology of China, Chengdu, China
e-mail: zhao_zhangzhi@sina.com

© Springer International Publishing AG 2018
H. Lu and X. Xu (eds.), *Artificial Intelligence and Robotics*,
Studies in Computational Intelligence 752,
https://doi.org/10.1007/978-3-319-69877-9_17

157

Based on the application scene of the intelligent stadium, the positioning of people or equipment is particularly important in the indoor environment. Global Positioning System (GPS) and Assisted Global Positioning System (A-GPS) have pushed outdoor positioning to the extreme, but many applications which are similar to large stadiums and other indoor places must have the ability to achieve seamless positioning in any environment. Although the indoor positioning system has made some progress in the past few decades, it is still the focus of global research. Therefore, in recent years, a large number of mobile applications based on location services (Location Based Services, LBSs) came into being, in which positioning system based on the intelligent stadium is necessary for research.

Indoor positioning technology based on the WIFI fingerprinting is a popular positioning method. Domestic and foreign research indicates that its positioning accuracy and robustness are better than triangulation and other methods based on the delay of radio frequency transmission [1]. In this project, we first study the positioning model of WIFI fingerprint positioning, and then research the method to improve the accuracy of WIFI fingerprint positioning, mainly to reduce the impact of equipment heterogeneity on the positioning accuracy. Simultaneously we use the built-in displacement sensor of smart phone to reduce the search space of fingerprint matching and improve the reliability of the location information [2–6] (Fig. 1).

The key to fingerprint positioning technology is to establish the relationship between fingerprint data and geographical location. And the signal strength of RSSI of WIFI obtained through the terminal equipment is the key to the establishment of fingerprint data. The various barriers are the most important part of the WIFI signal, and secondly, multipath propagation is another major factor affecting the WIFI signal.

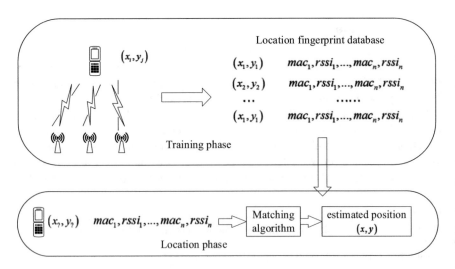

Fig. 1 WIFI fingerprint positioning principle diagram

In addition, the environmental factors (temperature, humidity, air flow velocity), WIFI network equipment and electromagnetic waves produced by various other electronic products will have a greater impact on the WIFI signal strength, resulting in the positioning accuracy of the positioning system.

Although there are many problems and shortcomings of WIFI fingerprint positioning due to above factors. However, when the measured points can accept more than five RSSI signal strength of AP, WIFI fingerprint positioning system can achieve a higher positioning accuracy, which can meet most of the positioning requirements [7]. Additionally, this method can utilize the APs that have been widely arranged and do not need the location information of the known APs, thus the positioning can be realized only using the WIFI information. Therefore, for the region where the current WIFI node has been arranged, the WIFI fingerprint location has a significant advantage over the location algorithm based on measuring distance [8].

2 WIFI Fingerprint Indoor Positioning Algorithm

2.1 WIFI Fingerprint Positioning Principle

The technology based on WIFI fingerprint positioning uses the related character-istics between spatial information and wireless signal RSSI, and matches the WIFI wireless information collected at the measured location with the geographic information. In the actual positioning environment, The RSSI values of the n APs received at the measured point form a set of n-dimensional vectors and constitute a mapping relationship with two-dimensional geographic locations. Different n-dimensional RSSI vectors correspond to different geographical locations, and then gather these n-dimensional RSSI. The vector constitutes a fingerprint database. Each set of data in the database is called a location fingerprint [9]. Finally, according to the change of the RSSI signal strength of the AP, the set of RSSI values measured at that time are uploaded to the location server. By matching with the position fingerprint of the fingerprint database, the geographical position cor-responding to the position fingerprint with the largest similarity is selected as the estimated position. The positioning algorithm based on WIFI fingerprint includes two stages: offline training phase and real-time positioning stage [10].

Offline training process: The main work of this stage is that the staff selects a number of reference points in the area to be measured and collects the signal intensity values from the different APs at each reference point. The RSSI value of AP received from each reference point, the MAC address, and the geo-location information of the reference point forms a set of associated ternary data sets stored in the database, where each set of data in the database is a location fingerprint [11].

Real-time positioning process: In the locating area, we collect the WIFI infor-mation of all the AP access points through the terminal device, and combine the

MAC address and the RSSI value of the AP into a binary group. The binary group is used as the input of the fingerprint positioning algorithm, and the closest entry is found by comparing one by one. Finally we use the nearest one or several sets of location fingerprints corresponding to the geographical coordinates to carry out the relevant calculations to estimate the location of the points to be measured.

2.2 WIFI Fingerprint Positioning Algorithm

The typical WIFI location fingerprint localization algorithm is divided into two types: the deterministic localization algorithm and the probabilistic localization algorithm [12]. The deterministic localization algorithm mainly includes: neighbor method, Artificial Neural Network and SVM algorithm. Probabilistic localization algorithm mainly includes Naive Bayesian. The nearest neighbor method includes: Nearest Neighbor, k Nearest Neighbor, weighted k Nearest Neighbor. In this paper, the WIFI location fingerprint localization algorithm is based on the weighted k Nearest Neighbor. Aiming at the problems and shortcomings of the proposed algorithm, this paper proposes a improved scheme of WKNN in order to achieve better positioning effect from the fingerprint database processing.

2.2.1 K-Means Clustering Method

The clustering is a process classifying data objects into one category. At present, there are a lot of algorithms which have been proposed to solve such problems. Among them, k-means algorithm, an analysis method dividing data set into k class, is one of the most popular clustering methods. In this paper, K-means is applied to the classification of fingerprint database, and the improved method is proposed to improve the efficiency of localization.

The core idea of the K-means algorithm can be explained as follows. First, we randomly select k center points from the n data, and calculate the distance of the remaining data to the center of each cluster. According to the calculated distance value, we select the nearest clustering center and return the data object to the cluster it represents. After that we calculate the average of all the data objects in each cluster and use it as the new center of each cluster. The iterative method is used to repeat the above process until the clustering center point is exactly the same as the last or converges to a given clustering criterion function, when the clustering result is optimal and the algorithm ends.

In the offline stage, WIFI Location Fingerprint location classifies the database according to the clustering algorithm. In the online positioning, the E-distance between signal strength value of RSSI measured at the sampling point and center of K clusters is first calculated. We find the nearest clustering center with the distance to be measured, and then match the point to the sample object of the sub-region where the cluster center is located to calculate the position coordinates, which can greatly improve the efficiency of March.

Steps:

(1) K clustering centers are arbitrarily selected from the data set X. $M = \{m_1^{(1)}, m_2^{(1)}, \ldots, m_k^{(1)}\}$

(2) Calculate the Euclidean distance of each remaining data $x_i (i = 1, 2, \ldots, N - K)$ to the K cluster centers M, denoted by $D(i, r)$. Assuming that n is the number of APs in the positioning area, and $x_i(RSSI_j)$ indicates the signal strength value of the j-th AP received by the i-th data object, and $m_r^{(1)}(RSSI_j)$ indicates that the r-th cluster center receives the signal strength value of the j-th AP. If the data object x_i has not take to the j-th AP signal, then replace with an infinitesimal numerical.As shown in Eq. 1.

$$D(i, r) = \sqrt{(x_i(RSSI_1) - m_r^{(1)}(RSSI_1))^2 + \cdots + (x_i(RSSI_n) - m_r^{(1)}(RSSI_n))^2} \quad (1)$$

(3) Assign x to the clustering center with the shortest European distance

(4) Calculate the average of all objects for each cluster and use it as a new clustering center point, as shown in Eq. 2.

$$m_r^{(2)} = \frac{1}{N_r} \sum_{t=1}^{N_r} x_t (x_t \in C_r, r = 1, 2 \ldots K) \quad (2)$$

In Eq. 2, C_r represents the data contained in class j, and N_r represents the number of j-th data.

(5) Repeat the above three steps, and classify the remaining x_i, when fluctuations of $M_r^{(p)}$ is less than the threshold of a given clustering criteria or cluster center no longer change, the output K class center, the algorithm ends.

Although K-means is a classical clustering algorithm, its clustering effect is very obvious when the distinction between class and class is large. However, the final clustering results of the algorithm are very sensitive to the initial clustering center., Different initial clustering centers selected leads to different clustering results. If the clustering center is badly chosen, the algorithm may fall into the local optimal solution, leading to the failure of the cluster. The usual practice is to divide the space K, and then select the initial cluster center in each region. Although the geographical location in some spaces is similar, the received RSSI signal strength values may be very different due to the layout and partition factors, resulting in poor clustering effect.

Therefore, this paper proposes to select the more suitable value K and clustering center according to the space and layout position of the actual area to be measured. The area can be divided into several areas with obvious spatial characteristics of the region, thus speeding up the matching efficiency and improving the positioning of real-time. This paper's processing flow of the WIFI location fingerprint positioning algorithm is as follows: In the offline stage, the position fingerprint database is established by measuring the RSSI signal value of APs to be measured and the fingerprint database is divided into K classes according to the above-mentioned method. In the online positioning stage, the signal intensity value of the AP received from the sampling point is taken as the sampling value. We first calculate the Euclidean distance from the center of K clustering, divide it into the nearest subclass, and then use the weighted K-close method to match the position and calculate the position coordinates.

3   Simulation Results Analysis

This paper uses the system to measure the intensity of the RSSI signal measured in the actual environment. The same device is used to test 20 points in the experimental environment. We compared the improved algorithm with WKNN algorithm that has not been improved. We can get the positioning error comparison chart from the positioning result error table, as shown in Fig. 2.

The analysis of test results: In the 20 test points, the maximum error of the traditional WKNN is 6.85 m and the minimum positioning error is 0.56 m, while the maximum positioning error of the improved WKNN is 5.83 m and the minimum positioning error is 0.50 m. Despite that the measurement error of the 11th group of data is the same, and the measurement error of the 4th and 8th groups is larger than that of the traditional WKNN, the rest 17 sets of data are obviously

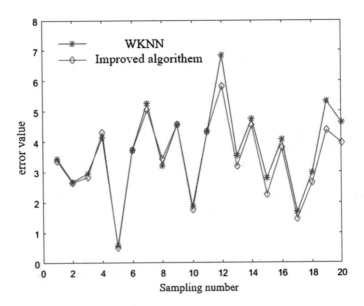

Fig. 2 Comparison chart of positioning error

Table 1 Location of the timetable

	Test time of WKNN positioning algorithm unit: milliseconds (ms)	Test time of improved algorithm unit: milliseconds (ms)
20 sets of test data for the total time	400	340
Average single positioning time	20	17

reduced. Results indicate that the improved algorithm has improved the accuracy of the traditional WKNN.

In addition, we also test the 20 sets of test data positioning time, including the overall system time-consuming and average time-consuming data per set, to compare the two algorithms of the algorithm complexity. In the same computer we run the two algorithm mentioned several times to complete the 20 sets of test data, and find the average of multiple measurement time to get the positioning time of the two algorithms. As shown in Table 1, it is important to note that the positioning time is related to the performance of the computer and the efficiency of the software. The results of the test are for reference only.

Analysis of test results: The total test time of the traditional WKNN is 400 ms, and the positioning time of the average single point is 20 ms. The improved algorithm reduces the amount of data in the total database due to the selection of AP. In the latter part of the algorithm, the clustering is also added, which makes the matching speed obviously and the performance of the system is optimized. Finally, the total test time is reduced by 60 ms and the average positioning time for a single point is reduced by 3 ms.

4 The Application of WIFI Positioning System in Intelligent Sports Venues

In this paper, the localization algorithm of WIFI positioning system in Wisdom Sports Stadium is based on the above optimized WIFI fingerprint algorithm and C/S (client/server) architecture. The server uses the development language for JAVA, and the fingerprint database uses the MySQL database, and the client uses the Socket to communicate. WIFI information processing, positioning algorithm implementation and WIFI signal information storage are done on the server side, so the server must achieve three major functions including database storage, Socket communication, and algorithm positioning. Through the above system architecture, we can achieve the positioning of people and objects, and make the terminal track the target smartly. As a result, we can complete identification and analysis of

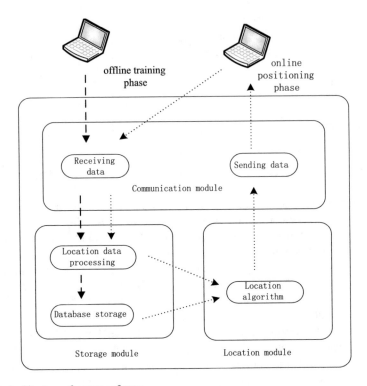

Fig. 3 Architecture of server software

Table 2 The database only designed a table Finger_Print, used to store location information, node information and its corresponding WIFI information

Column name	Description	Type	Restrictions
uuid	The id number assigned by the system	varchar (255)	Primary key
apid	The id of the point to be measured	int (11)	Not null
Positionx	X coordinate	float	Not null
Positiony	X coordinate	floatnull	Not null
Fingerdetail	WIFI information	longtext	Not null

multiple points of unmanned live and athlete action. Architecture of server software, as shown in Fig. 3.

The database only designed a table Finger_Print, used to store location information, node information and its corresponding WIFI information. Table 2 shows the structure of table.

5 Summary

Based on WIFI technology specification and its equipment, this paper proposes an improved fingerprint localization algorithm based on WKNN. The Experiments show that the algorithm proposed in this paper has higher accuracy and faster speed over WKNN. Based on the algorithm, this paper puts forward a solution of location service in intelligent sports venues. Positioning accuracy and data collection in intelligent stadium are the basis for the realization of various functions. In future research, we will innovate more ways to simplify the method of fingerprint collection, speed up the speed of data collection and data cleansing. Simultaneously, we will utilize the internal structures of the stadium to improve the positioning accuracy and practicality of navigation, combining with the method of machine learning and other research of information technology. We continue to iterate the wisdom of the stadium system, more intelligent and accurate to achieve live and analysis of athletes' action and dynamic management of the stadium to solve more difficult problems, which is a meaningful research work.

References

1. Dortz, N., Gain, F., Zetterberg, P.: WiFi fingerprint indoor positioning system using probability distribution comparison. In: 2012 IEEE International Conference on Acoustics, Speech and Signal Processing (ICASSP), pp. 2301–2304. IEEE (2012)
2. Chen, C.: Quickly Build WIFI Video Monitoring Car with Webcam, (004), pp. 5–7. Radio (2013)
3. Lu, H., Li, Y., Chen, M., Kim, H., Serikawa, S.: Brain intelligence: go beyond artificial intelligence. Mob. Netw. Appl. 1–10 (2017)
4. Lu, H., Li, Y., Mu, S., Wang, D., Kim, H., Serikawa, S.: Motor anomaly detection for unmanned aerial vehicles using reinforcement learning. IEEE Internet Things J. https://doi.org/10.1109/JIOT.2017.2737479
5. Lu, H., Li, B., Zhu, J., Li, Y., Li, Y., He, L., Li, J., Serikawa, S.: Wound intensity correction and segmentation with convolutional neural networks. Concurrency and Computation: Practice and Experience. https://doi.org/10.1002/cpe.3927
6. Serikawa, S., Lu, H.: Underwater image dehazing using joint trilateral filter. Comput. Electr. Eng. 40(1), 41–50 (2014)
7. Chen, J.H.: An application of robust template matching to user location on wireless infrastructure. In: Pattern Recognition, International Conference on IEEE Computer Society, p. 687690 (2004)
8. Heng, Y., Yaya, W., Bin, L., Dan Guo.: Positioning Technology, pp. 162–164. Electronic Industry Press, Peking (2013)
9. Mestre, P., Coutinho, L., Reigoto, L., et al.: Indoor location using fingerprinting and fuzzy logic. Adv. Intell. Soft Comput. 107, 363–374 (2011)
10. Xiang, Z., Song, S., Chen, J., et al.: A wireless LAN-based indoor positioning technology. IBM J. Res. Dev. 48(5), 617–626 (2004)

11. Robinson, M., Psaromiligkos, I.: Received signal strength based location estimation of a wireless LAN client. Wireless Communications and Networking Conference, 2005 IEEE, pp. 2350–2354. IEEE (2005)
12. Tran, Q., Tantra, J.: Wireless indoor positioning system with enhanced nearest neighbors in signal space algorithm. In: IEEE 64th Vehicular Technology Conference, pp. 1–5 (2006)

Force and Operation Analyses of Step-Climbing Wheel Mechanism by Axle Translation

Masaki Shiraishi, Takuma Idogawa and Geunho Lee

Abstract Obstacles such as ramps, steps and irregular floor surfaces are commonly encountered in homes, offices and other public spaces. These obstacles frequently limit the daily activities of people who use mobility aids. To expand the scope of their activities, a wheel mechanism for climbing a step while reducing the horizontal climbing force is proposed to be applied into self-propelled wheelchairs. The physical and mental burdens of caregivers and medical staff can be reduced by making the users of the mechanism more self-sufficient. Specifically, the proposed step-climbing mechanism for the self-propelled wheelchair relies on offsetting the rotational axis of the wheel for step climbing. This paper provides details on this offsetting mechanism and its force analyses.

Keywords Assist robots · Axial translation · Step-climbing · Wheelchair

1 Introduction

We often encounter thresholds, uneven surfaces and ramps in daily routines. People relying on wheel-type aids such as wheelchair [1] and walker [2] are often hindered by these obstacles, resulting in that their daily activities are limited. Many studies have been introduced to resolve these problems. For example, various auxiliary devices have been designed to attach to existing wheelchairs [3]. Other works have developed replacement parts with augmented step climbing capabilities. Some of these replacement wheels combine multiple smaller wheels into a larger wheeled

M. Shiraishi · T. Idogawa · G. Lee (✉)
Faculty of Engineering, University of Miyazaki, 1-1 Gakuen Kibanadai-nishi,
Miyazaki 889-2192, Japan
e-mail: geunho@cc.miyazaki-u.ac.jp

M. Shiraishi
e-mail: hi12025@student.miyazaki-u.ac.jp

T. Idogawa
e-mail: he11007@student.miyazaki-u.ac.jp

© Springer International Publishing AG 2018
H. Lu and X. Xu (eds.), *Artificial Intelligence and Robotics*,
Studies in Computational Intelligence 752,
https://doi.org/10.1007/978-3-319-69877-9_18

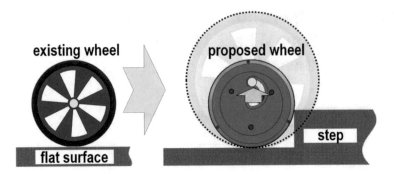

Fig. 1 Pseudo wheel diameter results from offsetting the rotation axis

assembly [4]. In the case of replacement parts, both the rear and front wheels have been considered for replacement and using crawlers instead of wheels has also been studied [5]. Moreover, some devices have been designed for step climbing from inception [6]. The devices from the studies suffer several drawbacks such as high cost, bulky size, heavy weight and high complexity. It is often difficult to integrate these devices into the existing low-cost assist devices.

In this paper, a wheel mechanism allowing to integrate into an existing assist device is proposed and realized. Specifically, our idea behind the wheel mechanism is to increase the rolling radius of a wheel by vertically offsetting its rotating axis from the wheel centre. As shown in Fig. 1, a pseudo wheel diameter results from this offsetting procedure. Accordingly, the wheel mechanism is capable of two types of motions. The first is normal rolling and the second is a separate motion reserved for step climbing. The step climbing motion aims to reduce the horizontal force required for step climbing. In this paper, the practical aspects such as horizontal forces and operations of the proposed mechanism are analyzed. The key contributions in our work are (1) realization of a step climbing unit and (2) which is capable of reliable step climbing motion. The wheel mechanism is expected to be applicable to various mobility aids as it is easily integrated and compact.

2 Wheel Mechanism for Rotational Axis

The key idea of the proposed solution for climbing a step is that the rolling radius of a wheel is increased by offsetting its rotational axis vertically (see Fig. 1). We expect that the axis offset will decrease the required horizontal force for step climbing and thereby reduce user effort. In addition, since wheeled devices are commonplace in mobility aids, results from this work could serve a wide audience including elderly and disabled people.

Wheeled transportation devices such as walkers, hand trucks and wheelchairs are commonly used in everyday life. These devices usually operate on flat surfaces and are adversely affected by irregularities such as steps, thresholds and dips. The ability

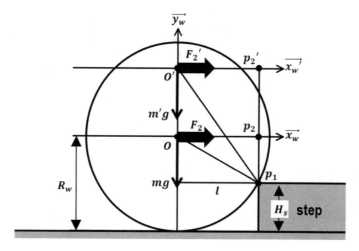

Fig. 2 Definition and notation for step climbing analysis

of a wheeled device to traverse such irregularities is dependent on the size of its wheels relative to the irregularity. In comparison to moving on a flat surface, a large force may be required to climb a step.

In this paper we consider the forces required for rolling a wheel on a flat surface and for climbing a step. It is assumed that the step height is smaller than the wheel radius. As shown in Fig. 2, the rolling direction of the wheel is defined as $\mathbf{x}_w$ with its origin at O, the axis given by a $90°$ counter-clockwise rotation of $\mathbf{x}_w$ about O is $\mathbf{y}_w$, the point where the wheel contacts the step is denoted as p_1 and its normal projection onto $\mathbf{x}_w$ is denoted by p_2. Moreover, the wheel radius is defined as R_w, the wheel mass as m, gravitational acceleration as g and the coefficient of rolling friction as μ, respectively. First, the force required to move on a flat surface is denoted by F_1. Secondly, we calculate the required force to climb a step. The step climb is accomplished by virtue of a clockwise moment $M_2 (= |\overline{p_1p_2}| \cdot F_2)$ which results from the force couple between the applied external force at O and the contact point reaction force at p_1. Here, the distance from p_1 to p_2 is denoted by $|\overline{p_1p_2}|$. The horizontal distance between p_1 and O is denoted by l, as shown in Fig. 2. The height of the step is defined as H_s, so that l is obtained as $l = \sqrt{R_w^2 - |\overline{p_1p_2}|^2}$ by the Pythagorean theorem. In order to climb a step, the clockwise moment due to the external force must be greater than the counter-clockwise moment due to the weight of the wheel, resulting in $|\overline{p_1p_2}| \cdot F_2 \geq l \cdot mg$. Based on this conditional expression, the condition for F_2 can be derived:

$$|\overline{p_1p_2}| \cdot F_2 \geq \sqrt{R_w^2 - |\overline{p_1p_2}|^2} \cdot mg$$
$$\Longrightarrow F_2 \geq \frac{mg}{|\overline{p_1p_2}|} \sqrt{R_w^2 - |\overline{p_1p_2}|^2}. \tag{1}$$

The wheel diameter and mass is assumed to be 100 mm and 5 kg respectively. From (1), for step heights of 10, 20, 30 and 40 mm, the required forces are found to be 36.8, 65.3, 112.3 and 240.0 N respectively. By the these calculations, it is confirmed that the required force F_2 increases with the height of the step. Next, for design purposes, wheel diameters of 100 and 200 mm are compared in terms of the required horizontal force for step climbing. By assuming a step height of 10 mm and using (1), F_2 is found to be 36.75 and 18.85 N respectively.

3   Mechanism and Analysis of Axle Translation

This section introduces the rotation axis translation and horizontal force required to climb a step. Figure 3 illustrates the procedure of traversing a step is described.

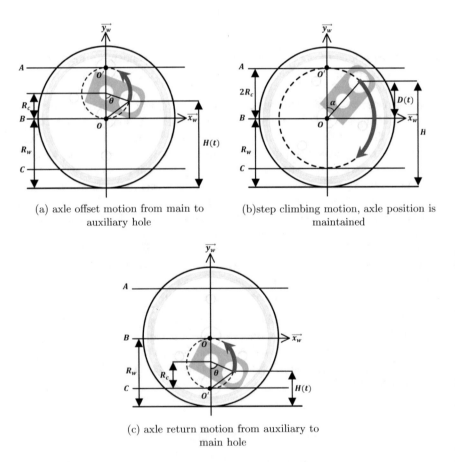

(a) axle offset motion from main to auxiliary hole

(b) step climbing motion, axle position is maintained

(c) axle return motion from auxiliary to main hole

Fig. 3   Axle position offset, maintenance and return for the step climbing procedure

In consecutive order, the rotation axis of the wheel is translated vertically, a step is traversed, and the rotation axis of the wheel is returned to its original position. Figure 3a, c show the mechanism by which the rotation axis of the wheel is offset and returned respectively. To clarify the explanation, the motion is divided into three separate motions with distinct time phases $(k_0 \leq t < k_1)$, $(k_1 \leq t \leq k_2)$ and $(k_2 < t \leq k_3)$. We define the crank radius of rotation as R_c and the height of the axle relative to the bottom of the wheel at each time t as $H(t)$. Here, $H(t)$ varies depending on the rotation of the crank and the wheel. In particular, the position where the axle is at its highest relative to the bottom of the wheel is defined as A, the position where the axle is centred at the wheel centre is defined as B and the position where the axle is at its lowest relative to the bottom of the wheel is defined as C.

First, directly in front of the step, the axle is detached from the main hole of the wheel, and the crank is rotated so that the end of the axle aligns with the auxiliary hole. The axle then attaches to the auxiliary hole. During this procedure the rotation axis of the wheel is translated vertically from O to O' as shown in Fig. 3a. For $(k_0 \leq t < k_1)$, $H(t)$ is derived as function of the rotation angle of the crank $\theta(t)$:

$$H(t) = R_w + R_c - R_c \cos \theta(t) \tag{2}$$

where, $H(t)$ is at its minimum value B when $t = k_0$, similarly, its maximum value A is obtained when $t = k_1$.

Secondly, we discuss the rolling of the wheel when it is climbing a step. To traverse a step, the wheel rotates about two distinct axes: centred at the main hole and auxiliary hole respectively. During the climbing phase, we consider the case where the wheel is rotating about the axis centred at the auxiliary hole. When $(k_1 \leq t \leq k_2)$, $D(t)$ is given by $D(t) = 2R_c \cos \alpha(t)$ which is obtained by using the angle $\alpha(t)$ between $\mathbf{y_w}$ and the straight line connecting the centre of the wheel and the axis of rotation of the wheel as shown in Fig. 3b. The height of axle above the ground is $H(t) = R_w + D(t)$. By substitution of $D(t)$ the following equation is obtained:

$$H(t) = R_w + 2R_c \cos \alpha(t) \tag{3}$$

where, $H(t)$ is at its maximum value at A when $t = k_1$. Once $H(t)$ has been moved to A, the climb is performed. After the climb, $H(t)$ is at its minimum value at the height of the C when $t = k_2$.

Thirdly, we discuss the vertical translation of the wheel rotation axis by the crank after the climbing phase. Once again, directly after the climbing phase is complete, the axle is detached from the auxiliary hole of the wheel, and the crank is rotated so that the end of the axle aligns with the main hole. The axle then attaches to the main hole. During this procedure the rotation axis of the wheel is returned from O' to O via vertical translation as shown in Fig. 3c. Since the axle is now centred at the wheel centre, forward motion can be resumed on the flat surface. When $(k_2 < t \leq k_3)$ in Fig. 3c, $H(t)$ is derived:

$$H(t) = R_w - R_c - R_c \cos \theta(t) \tag{4}$$

where $H(t)$ is at its minimum value at the height of the C when $t = k_2$. Moreover, $H(t)$ is maximum when $t = k_3$ and the height of the axle relative to the bottom of the wheel becomes B (the same as before the climbing phase).

4   Results and Discussion

In this section, we investigated the required step climbing force applied to the axle for various cases. In Sect. 2, the horizontal force required for step climbing was analysed. The horizontal force was applied to the wheel from the axle centre. In practice, the required horizontal force is generated by a torque applied to the rear wheel of the wheelchair. This force is transferred to the front axle through the wheelchair frame. The required torque was determined by incrementally increasing the applied torque until the front wheel cleared the step. Moreover, the required torque for step climbing was determined for 10 mm increments in the step height. Figure 4 shows the required force for three different cases, calculated from (1) and (4). The results for the same cases are also plotted in Fig. 4 for comparison. A good correlation is found between the analytical and experimental results. This effectively validates the correctness of the derived equations.

· Case #1:   160 mm wheel diameter, the height of the axis above the ground is 80 mm
· Case #2:   160 mm wheel diameter, the height of the axis above the ground is 130 mm
· Case #3:   260 mm wheel diameter, height of the axis above the ground is 130 mm

By comparison of Cases #1 and #3, the step climbing force is smaller for Case #3. Additionally, the wheel in Case #3 is able to climb higher steps because of its axle height. By comparing Cases #1 and #2, the step climbing force in Case #2 is smaller. The wheel in Case #2 can climb higher steps due to its axle height. In addition, comparison of Cases #2 and #3 shows that the step climbing force in Case #2 is smaller. The smallest step climbing force is obtained in Case #2 for all step heights. Moreover, this comparison of the three cases also shows that the difference in required force between the cases increases as the step height increases. The results effectively show that it is possible to realize a small step climbing force without requiring a large wheel diameter.

The wheel mechanism as explained in Sect. 4 causes a change in the rolling radius by changing the position of the axle. The proposed axle translation motion leads to a small required step climbing force according to (1) and (6). As a result, the wheel can climb the step up to about 31% of its diameter when the axle is in the wheel centre. However, the wheel can climb up to about 81% of its diameter when the axle translation motion is carried out. Under the current mathematical model, it is possible for a wheel to climb up to $2R_w$. However, the developed prototype is subject to manufacturability constraints, as a result the maximum climbable step

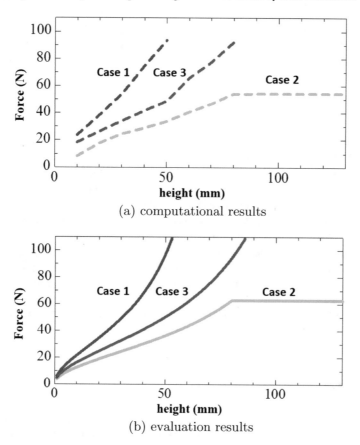

(a) computational results

(b) evaluation results

Fig. 4 Analytical results required horizontal forces

height is $1.8R_w$ for the prototype. By using the proposed mechanism, it is possible to perform step climbs higher than R_w, which is not possible with a conventional wheel. In addition, the required climbing force is much lower with a small wheel and the proposed mechanism, than for a comparatively large conventional wheel. Therefore, it is confirmed that the wheel's potential to climb a step is improved by the proposed mechanism.

The wheel mechanism is designed for use in wheelchairs and hand trucks. The prototype is integrated into a commercial self-propelled wheelchair. Due to the page limitation, the hardware configuration and the implementation of the mechanism are omitted in this paper. Details on the realizations can be in [7]. In practical, the prototype weighs about 5 kg. It is possible to design a lighter and more compact system by the choice of materials and off-the-shelf parts. The current wheel mechanism can only climb the step directly from the front. It is possible to design a mechanism that allows for climbing the step from varying directions. In this paper, a wheelchair was used for conducting the experiments. On the wheelchair the diameters of the front and rear wheels are different. Therefore, the rear wheel climbing motion needs to be considered. In future the safety aspects of using the wheel mechanism also need to be considered.

5 Conclusions

In this paper, a wheel mechanism for step climbing is proposed. The proposed mechanism vertically offsets the rotational axis of the wheel and thereby increases its rolling radius. Thus, it is possible to increase the pseudo wheel diameter for step climbing. To prove the effectiveness and practicality of the wheel mechanism, step climbing experiments were carried out for three different cases. It is shown that the proposed system can outperform a conventional wheel with larger diameter in terms of required step climbing force. The proposed mechanism therefore offers a more compact system and also reduces user effort. Future research directions include weight reduction, consideration of safety and expansion of use range [8].

Acknowledgements This work is partially supported by the Research Grant-A awarded by the Tateisi Science and Technology Foundation.

References

1. Carlson, T., Leeb, R., Chavarriaga, R., Millan, J.R.: The birth of the brain-controlled wheelchair. In: Proceedings of the IEEE/RSJ International Conference on Intelligent Robots and Systems, pp. 5444–5445 (2012)
2. Lee, G., Ohnuma, T., Chong, N.Y., Lee, S.-G.: Walking intent based movement control for JAIST active robotic walker. IEEE Trans. Syst. Man Cybern. Syst. **44**(5), 665–672 (2014)
3. Mori, Y., Katsumura, K., Nagase, K.: Development of a pair of step-climbing units for a manual wheelchair user. Trans. JSME **80**(820) (2010). https://doi.org/10.1299/transjsme.2014dr0381 (in Japanese)
4. Quaglia, G., Franco, W., Oderio, R.: Wheelchair.q, a motorized wheelchair with stair climbing ability. Mech. Mach. Theor. **46**(11), 1601–1609 (2011)
5. Razak, S.: Design and implementation electrical wheel chair for disable able to stairs climbing by using hydraulic jack. IOSR J. Electr. Electron. Eng. **7**(3), 82–92 (2013)
6. Sasaki, K., Eguchi, Y., Suzuki, K.: Step-climbing wheelchair with lever propelled rotary legs. In: Proceedings of the IEEE/RSJ International Conference on Intelligent Robots and Systems, pp. 1190–1195 (2015)
7. Shiraishi, M., Lee, G., Idogawa, T.: Axle translation based step-climbing wheel mechanism for existing passive assist devices. In: Proceedings of the International Conference on Design and Concurrent Engineering (2017)
8. Jager, P.: Cebrennus Simon, 1880 (Araneae: Sparassidae): a revisionary up-date with the description of four new species and an updated identification key for all species. Zootaxa **3790**(2) (2014). https://doi.org/10.11646/zootaxa.3790.2.4

A Policing Resource Allocation Method for Cooperative Security Game

Zhou Yang, Zeyu Zhu, Ling Teng, Jiajie Xu and Junwu Zhu

Abstract How to make good use of different kinds of security resources to protect the urban city has always been an important social problem. Firstly, we propose a multi-agent framework in which various kinds of agents cooperate to complete tasks. In this framework, we develop an auction-based task allocation mechanism for police force agents. Secondly we come up with a double oracle algorithm for policing resources allocation in cooperative security games on graphs. Lastly, according to the experimental results, our algorithm runs approximately 25% faster than the former work.

Keywords Game theory · Double oracle · Distributed artificial intelligence
Multi-agent system · Task allocation

1 Introduction

Security has always been a gigantic challenge faced by the world. According to the data, 13,463 terrorism accidents took place in 2014, killing more than 32,700 people and injuring more than 34,700 people. Many cities have applied different kinds of security measures to protect the potential attack targets, e.g., shopping centers, tourist sites, schools, subway stations.

Recently, Multi-Agent System has become a research hotspot in Distributed artificial intelligence field. It is also significantly useful in security domains. For example, in urban cities with changing environments and extremely complicated

Z. Yang · L. Teng · J. Xu · J. Zhu (✉)
College of Information Engineering, Yangzhou University, Yangzhou, China
e-mail: jwzhu@yzu.edu.cn

Z. Zhu
School of Computer and Information Engineering, Guangxi Teachers Education University,
Nanning, China

J. Zhu
Department of Computer Science, University of Guelph, Guelph, ON, Canada

© Springer International Publishing AG 2018
H. Lu and X. Xu (eds.), *Artificial Intelligence and Robotics*,
Studies in Computational Intelligence 752,
https://doi.org/10.1007/978-3-319-69877-9_19

tasks, it is hard for single agent to completed rescuing tasks. This motivates us to develop a multi-agent framework in which different agents cooperate to complete the task together. In this framework, there are various kinds of heterogeneous agents with different features and abilities.

In this paper, we mainly concentrate on police force agents. Auction is a fast and efficient resource allocation scheme with strong operability. It can help the auctioneers and the bidders both achieve ideal utilities. We propose an auction mechanism to allocate tasks for police force agents.

Game theory provides an appropriate mathematical model for researching optimal allocating strategies of limited resources. Stackelberg games were first introduced to model leadership and commitment [1]. It corresponds the situation faced by security agencies in security domains. We develop a double oracle based algorithm for allocating police resources in the city road network.

In Sect. 2, we describe the problem and introduce our system framework. In Sect. 3, we mainly discuss the double oracle based policing resource allocation method. Then we show the experiment result and do some analyze. In the Sect. 5, we show some related work and draw some conclusion.

2 System Framework and Problem Description

2.1 System Framework

We introduce the multi-agent framework in the cooperative security game.

According to the moving ability, agents in our system framework can be divided into two kinds: the moving agents and the static agents. Moving agents are agents that can move in the road network and can perform specific action, for example, fire brigades. The static agents are always center agents, like fire station. They can not move but can give orders and direct corresponding agents.

According to the functions, these agents can be divided into more kinds, police force agents for patrolling, ambulance team agents for rescuing civilians, fire brigade agents for putting out fires and other agents for other various tasks. All these agents are leaded by their respective center agents.

In this paper, we mainly focus on the police agents. So we ignore many details about other agents. Figure 1 is the system framework figure. In the circle marked with red-dotted lines are the static agents. In the circle marked with green-dotted lines are the moving agents. The solid bidirectional arrow between the ambulance station and the police station means that the two station can communicate with each other beyond the distance limitation. The bidirectional arrow with dot lines between moving agents means that the they can communicate with each other beyond the distance limitation. All the agents need to cooperate together to solve security problems.

When urgent issues emerge, police force agents will be assigned to work them out. We develop an auction-based task allocation mechanism, whose flow chart is shown in Fig. 2.

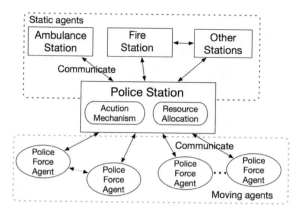

Fig. 1 The system framework figure

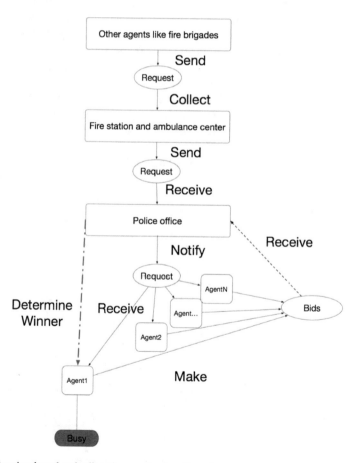

Fig. 2 Auction-based task allocation mechanism flow chart

When agents besides police forces find emergency issues in the city, they will send requests to their corresponding center agents. Those center agents collect requests and send them to police office. Police office notifies all the free police force agents about those requests (tasks). Then all the police agents make bids on tasks. After receiving bids, police office determines the winner and the winning agent goes to solve the issue. Other police force agents are set as free state. They are allocated to patrol in some places in the city map. Such resources allocation problem is described in the next sub-section.

2.2 Resource Allocation Problem Description

We model the policing resource allocation problem as a zero-sum game between the defender (police force agents) and the potential attacker. The pure strategies of the attacker are all the possible continuous sequences of nodes from any source $s \in s \subset N$ to any target $t \in T \subset N$. We assume that the defender plays the allocation $X_i \subseteq X$ and the attacker chooses path $A_j \subseteq A$. The attacker aims to reach targets safely and the defender tries to stop the attacker before his arrival.

The attacker succeeds if and only if he is not detected all alone his path. The attacker gets $\tau(t)$ for a successful attack on t and 0 otherwise. The objective is to find a mixed strategy x for the defender which can maximize the expected utility of the defender, given that the attacker learns the mixed strategy of the defender and chooses a best-response for himself.

The notations used in this paper are defined in Table 1.

Table 1 The definitions of notations used in this paper

Notations	Meanings
$G = (N, E)$	A network graph
$\tau(t_j)$	The payoff of target t_j
k	The number of defender resources available
X	Set of defender allocations, $X = \{X_1, X_2, \ldots, X_n\}$
X_i	ith defender allocation. $X_i = \{X_{ie}\} \forall e, X_{ie} \in \{0, 1\}$
A	Set of attacker path. $A = \{A_1, A_2, \ldots, A_m\}$
A_j	jth attacker path. $A_j = \{A_{je}\} \forall e, A_{je} \in \{0, 1\}$
x	Defender's mixed strategy over X
a	Attacker's mixed strategy over A
p_n	Detection probability of node n
$U_d(x, A_j)$	Defender's expected utility playing x over A_j
X^*	Defender's pure strategy best response
A^*	Attacker's pure strategy best response

3 Double Oracle Approach for Finding the Optimal Strategy

In this section, we present DOAN (a Double-Oracle based Algorithm for allocating resources on Nodes) in graph-based security games. The double oracle algorithm aims to create a restricted game, where the players have limited numbers of allowed allocations. The the game is expanded iteratively by adding best-response strategies of both sides to strategy spaces.

Shown below, X and A are strategies generated so far in the current games. They increase along with the development of DOAN. $CoreLP(X,A)$ aims to find an equilibrium (x, a) of the current restricted games. It returns the mixed strategies of both sides at the equilibrium. Double oracle in this algorithm means defender oracle and the attacker oracle, which return a pure strategy of both sides. DO returns a best response strategy X^* for the defender against a. AO returns a best response strategy A^* for the attacker against x. However, DO is a NP-hard problem. So we develop a greedy algorithm $DO'(a)$ for the defender to find better responses which may not the optimal but run much faster.

In line 4, we call the function $DO'(a)$. It also returns a response strategy X^* for the defender against a. In line 5, if $DO'(a)$ can't provide a better pure strategy, we call $DO(a)$. By this way, we can decrease the numbers of calling $DO(a)$, improving the efficiency of the algorithm sharply. We do the same thing on the attacker oracle.

Algorithm 1 DOAN
1.   Initialize X and A
2.   **Do**
3. $(x, a) \leftarrow CoreLP(X, A)$
4. $X^* \leftarrow DO'(a)$
5.   if $U_d(X^*, a) < U_d(x, a)$ then
6. $X^* \leftarrow DO(a)$
7. $X \leftarrow X \cup \{X^*\}$
8. $X^* \leftarrow DO'(a)$
9.   if $U_d(X^*, a) < U_d(x, a)$ then
10. $A^* \leftarrow AO(a)$
11. $A \leftarrow A \cup \{A^*\}$
12.   while $U_d(x, A^*) \leq U_d(x, a)$ or $U_d(x, a) \leq U_d(X^*, a)$
13.   return (x, a)

Algorithm 1. Main algorithm showing the main procedure of DOAN.

One of the utility function can be formulated as below:

$$U_d\left(x, A^*\right) = -\tau\left(t_j\right) \cdot \sum_i \left(x_i \cdot \prod_{n \in A^* \cap X_i} \left(1 - p_n\right)\right) \tag{1}$$

The algorithm will stop only when $U_d(x, A^*) \leq U_d(x, a)$ **or** $U_d(x, a) \leq U_d(X^*, a)$ is not satisfied. If both oracles can not generate strategies providing better expected utility, the algorithm converges. For the defender, it means $U_d(x, a) \leq U_d(X^*, a)$. For the attacker, it means $U_a(x, A^*) \geq U_a(x, a)$. Since it is a zero-sum game, $U_a = -U_d$. So $U_a(x, A^*) \geq U_a(x, a)$ can be represented as $U_d(x, A^*) \leq U_d(x, a)$. The algorithm terminates when both conditions are not satisfied.

The correctness of best-response-based double oracle algorithm for two-player zero-sum games has been established by McMahan et al. [2]. In addition, we assert that the algorithm must converge or it must jump out of while-module. Because in the worst situation, it will generate all the pure strategies.

4 Experimental Results and Analysis

In this part we do experiment with DOAN. We introduce the experiment settings in section first and then analyze the experimental results in detail.

We do stimulations with DOAN. The experiments were conducted on the undirected graphs generated randomly. We mainly discuss the runtime results which represents performance of the algorithm. We mainly conducted experiments on weakly fully connected graphs.

We do experiments under different data size settings. Table 2 represents the running time of our algorithm in different situations.

Row represents sizes of graphs and column represents the number of resources available. As demonstrated above, (300,4) costs just a little bit more than half a minute. Compared with the results from former work [3], the time cost of our algorithm decreases about 25%.

Table 2 Running time of the algorithm with greedy algorithms	1	2	3	4
100	1.32	1.33	7.62	8.84
200	1.90	2.23	10.51	21.86
300	3.42	3.77	17.45	35.63

5   Related Work and Conclusion

Artificial intelligence (AI) is an important technology that supports daily social life and economic activities [4–7]. As an important tool of this field, game theory is a perfect model for allocating limited resources efficiently. The security game theory models the interaction between the defender and the attacker as Stackelberg games, which was proposed in 1930s [1]. Then, Nash and Stackelberg strategies were introduced into intelligent control and optimization by Wang [8]. In 2006, Conitzer published the fundamental paper [9] applying Stackelberg games to optimize the allocation for limited resources. The term Stackelberg Security Games was first introduced by Kiekintveld et al. [10] to describe security games. Many system have been developed for practical usage. ARMOR [11] help randomize checkpoints on the roadways entering the airport. IRIS [12] system aims to schedule air policemen for planes. PROTECT [13] helps the US Coast Guard design patrol routes. PAWS [14] was designed for protecting the wild animals from being hunting.

The techniques used in our model are based on a double oracle approach, proposed by McMahan et al. [15]. Double oracle algorithms have subsequently been applied to various pursuit-evasion games [16, 17]. Jain et al. [3] came up with the double oracle algorithm called RUGGED. Proved by [3], the best-response oracle problems for allocating resources on edges are NP-hard, new algorithm [18] has been propose to decreasing running time.

In this paper, a multi-agent framework in cooperative security games is introduced. We propose an auction-based task allocation mechanism police agents. We also design a double oracle approach for allocating policing resources on nodes. It can be solved in acceptable time cost (almost half a minute in a graph with 300 nodes and 4 resources).

Acknowledgements Project supported by the National Nature Science Foundation of China (Grant No. 61170201, No. 61070133, No. 61472344), Six talent peaks project in Jiangsu Province (Grant No. 2011–DZXX–032), Jiangsu Science and Technology Project (Grant No. BY2015061-06, BY2015061-08), Yangzhou Science and Technology Project (Grant No. SXT20140048, SXT20150014, SXT201510013), Natural Science and Technology Project (Grant No. SXT20140048, SXT20150014, SXT201510013), Natural Science Foundation of the Jiangsu Higher Education Institutions (Grant No. 14KJB520041).

References

1. von Stackelberg, H.: Marktform und Gleichgewicht. Springer, Vienna (1934)
2. Yin, Z., Korzhyk, D., Kiekintveld, C., Conitzer, V., Tambe, M.: Stackelberg vs Nash in security games: interchangeability, equivalence, and uniqueness. In: AAMAS, pp. 1139–1146 (2010)
3. Jain, M., Korzhyk, D., Van, et al.: A double oracle algorithm for zero-sum security games on graphs. In: International Conference on Autonomous Agents and Multiagent Systems. DBLP, pp. 327–334 (2011)
4. Lu, H., Li, Y., Chen, M., et al.: Brain intelligence: go beyond artificial intelligence (2017)

5. Lu, H., Li, Y., Chen, M., Kim, H., Serikawa, S.: Brain Intelligence: Go beyond artificial intelligence. Mob. Netw. Appl. 1–10 (2017)
6. Lu, H., Li, Y., Mu, S., Wang, D., Kim, H., Serikawa, S.: "Motor anomaly detection for unmanned aerial vehicles using reinforcement learning". IEEE J. Int. Things https://doi.org/10.1109/JIOT.2017.2737479
7. Serikawa, S., Lu, H.: underwater image dehazing using joint trilateral filter. Comput. Electr. Eng. 40(1), 41–50 (2014)
8. Wang, F.Y., Saridis, G.N.: A coordination theory for intelligent machines. Automatica 26(5), 833–844 (1990)
9. Conitzer, V., Sandholm, T.: Computing the optimal strategy to commit to. In: ACM Conference on Electronic Commerce DBLP, pp.82–90 (2006)
10. Kiekintveld, C., Jain, M., Tsai, J., Pita, J., Tambe, M., Ordonez, F.: Computing optimal randomized resource allocations for massive security games. In: International Conference on Autonomous Agents and Multiagent Systems (AAMAS), pp. 689–696 (2009)
11. Pita, J., Jain, M., Ordóñez, F., et al.: Using game theory for Los Angeles Airport Security. Ai Mag. 30(1), 43–57 (2009)
12. Tsai, J., Rathi, S., Kiekintveld, C., et al.: IRIS-a tool for strategic security allocation in transportation networks. In: Adaptive Agents and Multi-Agents Systems (2009)
13. Shieh, E., An, B., Yang, R., et al.: Protect: A deployed game theoretic system to protect the ports of the United States. In: International Conference on Autonomous Agents and Multiagent Systems, pp. 13–20 (2012)
14. Yang, R., Ford, B., Tambe, M., et al.: Adaptive resource allocation for wildlife protection against illegal poachers. In: International Conference on Autonomous Agents and Multiagent Systems (2014)
15. Mcmahan, H.B., Gordon, G.J., Blum, A.: Planning in the presence of cost functions controlled by an adversary. In: Proceedings of the International Conference on Machine Learning (ICML), pp. 536–543 (2003)
16. Halvorson, E., Conitzer, V., Parr, R.: Multi-step multi-sensor hider-seeker games. In: International Joint Conference on Artifical Intelligence, pp. 159–166. Morgan Kaufmann Publishers Inc. (2009)
17. Vaněk, O., Bošanský, B., Jakob, M., et al.: Transiting areas patrolled by a mobile adversary. In: IEEE Computational Intelligence and Games, pp. 9–16 (2010)
18. Jain, M., Conitzer, V., Tambe, M.: Security scheduling for real-world networks. In: International Conference on Autonomous Agents and Multi-Agent Systems, pp. 215–222 (2013)
19. Yang, R., Fang, F., Jiang, A.X., et al.: Designing better strategies against human adversaries in network security games. In: International Conference on Autonomous Agents and Multiagent Systems, pp. 1299–1300 (2012)

Global Calibration of Multi-camera Measurement System from Non-overlapping Views

Tianlong Yang, Qiancheng Zhao, Quan Zhou and Dongzhao Huang

Abstract Global calibration has direct influence on measurement accuracy of multi-camera system. The present calibration methods are hard to be applied in field calibration for the usage of complicated structures with accurate geometric or simple parts requiring overlapping view field of the system. Aiming at these problems, a new method is proposed in this paper by using two fixed plane targets with invariable pose. Objective functions are established according to constantness of the distance between original points and the axis angles of the plane targets, and nonlinear optimization is improved by means of Rodrigues transform. An apparatus is manufactured for real calibration experiments, and results verify the effectively and reliability of the method.

Keywords Multi-camera · Calibration · Rodrigues · Measurement
Plane target

1 Introduction

Monocular vision measurement system always contains two or more cameras for expanding measurement field, global calibration is a necessary process to unify measurement data provided by single camera for the independence of each camera

T. Yang · Q. Zhao (✉) · Q. Zhou
College of Mechanical and Electrical Engineering, Hunan University of Science
and Technology, Xiangtan, China
e-mail: qczhao@163.com

T. Yang
e-mail: eastlife0108@163.com

Q. Zhou
e-mail: zhouquan1103@163.com

D. Huang
Hunan Provincial Key Laboratory of Health Maintenance for Mechanical Equipment,
Hunan University of Science and Technology, Xiangtan, China
e-mail: husthdz@163.com

© Springer International Publishing AG 2018
H. Lu and X. Xu (eds.), *Artificial Intelligence and Robotics*,
Studies in Computational Intelligence 752,
https://doi.org/10.1007/978-3-319-69877-9_20

coordinate system, and it has direct influence on measurement accuracy of the system. Currently, a few methods have been in use for the calibration, there are two typical types [1–4, 8–10]. One of them uses a simple structure, there is no need to be measured beforehand, but it requires overlapping views of the system. The other one set up a structure in the separated views, in advance of the calibration, the structure geometric should be measured with coordinate measuring machine or other measuring equipment. Thus, the present methods are hard to be applied in field calibration.

In this paper, a new method and its associated apparatus are proposed for the global calibration. The apparatus contains two fixed planar targets. In process of the calibration, we change the position of the apparatus for presenting two targets with different poses in the cameras, relationship between the cameras can be calculated out based on constantness of the distance between original points and the axis angles of two targets. Benefited from the proposed method, it isn't necessary to use a part accurately measured beforehand, and it is appropriate for the system without any overlapping view.

2 Basic Notations

A 3D point is denoted by $\mathbf{P} = [x\ y\ z]^T$, its homogeneous coordinates are written by $\mathbf{P}^* = [x\ y\ z\ 1]^T$. A usual pinhole camera model is used in this paper, Wang's method [5] is adopted for obtaining intrinsic parameters of the camera. In this paper, targets are designed with chequerboard pattern, Lars Krüger's method [6] is used for corner detection. PnP solution [7] is used for external matrix calculation. Let $\mathbf{M}$ be homogeneous expression of the external matrix:

$$\mathbf{M} = \begin{bmatrix} \alpha & \beta & \gamma & \mathbf{t} \\ 0 & 0 & 0 & 1 \end{bmatrix} \tag{1}$$

where, α, β and γ can be identified as the vectors of three coordinate axes. $\mathbf{t}$ can be determined as the origin coordinates.

A 3×3 rotation matrix is denoted by $\mathbf{R}(\mathbf{k}, \phi)$, ϕ means the rotation angle, $\mathbf{k}$ means the unit rotation vector. $\mathbf{R}(\mathbf{k}, \phi)$ can be expressed in the form of Rodrigues:

$$\mathbf{R}(\mathbf{k}, \varphi) = \cos \varphi \mathbf{I} + (1 - \cos \varphi)\mathbf{k}\mathbf{k}^T + \sin \varphi \mathbf{k}^\times \tag{2}$$

where $\mathbf{I}$ is a 3×3 unit matrix, $\mathbf{k}^\times$ is a matrix defined by $\mathbf{k}$:

$$\mathbf{k} = \begin{bmatrix} k_x \\ k_y \\ k_z \end{bmatrix} \quad \mathbf{k}^\times = \begin{bmatrix} 0 & -k_z & k_y \\ k_z & 0 & -k_x \\ -k_y & k_x & 0 \end{bmatrix} \tag{3}$$

We denote the rotation of **P** by **RP**, let $\Delta\boldsymbol{\psi} = \Delta\phi\mathbf{k}$, $\Delta\mathbf{R} = \mathbf{R}(\mathbf{k}, \Delta\phi)$, suppose $\Delta\phi$ is an infinitesimal element, from Eq. (2, 3) a proximate equation is established:

$$\Delta\mathbf{R} = \mathbf{I} + (\Delta\boldsymbol{\psi})^{\times} \tag{4}$$

Let $\Delta(\mathbf{RP}) = (\Delta\mathbf{R})\mathbf{RP} - \mathbf{RP}$, from Eq. (2, 4) we have

$$\Delta(\mathbf{RP}) = (\Delta\mathbf{R} - \mathbf{I})\mathbf{RP} = (\Delta\boldsymbol{\psi})^{\times}\mathbf{RP} = -(\mathbf{RP})^{\times}\Delta\boldsymbol{\psi} \tag{5}$$

Equation (5) can be deduced as follow:

$$\frac{\partial(\mathbf{RP})}{\partial\boldsymbol{\psi}} = -(\mathbf{RP})^{\times} \tag{6}$$

3 Calibration Principle

3.1 Constraints

As shown in Fig. 1, two cameras with separated view field are taken as a sample to describe the principle. Let's denote two panel target by Target 1 and Target 2, two cameras by Camera 1 and Camera 2, Target 1 and 2 are presented in visible areas of Camera 1 and 2, respectively. C_{c1} and C_{c2} are camera coordinate systems and C_{t1} and C_{t2} are target coordinate systems. Transform matrix of two cameras is denoted by $\mathbf{T}_c$, which consists of a 3 × 3 matrix and a 3 × 1 vector:

$$\mathbf{T}_c = \begin{bmatrix} \mathbf{r}_{c21} & \mathbf{t}_{c21} \\ \mathbf{0} & 1 \end{bmatrix} \tag{7}$$

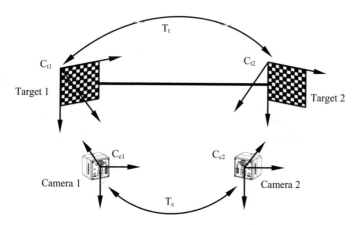

Fig. 1 Scheme of calibration principle

3.2   Rough Estimation

The apparatus is set up in front of a camera and two targets are visible in effective view for rough estimation. Let $\mathbf{M}_1$ and $\mathbf{M}_2$ be the external matrices of Target 1 and 2 in the camera:

$$\mathbf{M}_1 = \begin{bmatrix} \boldsymbol{\alpha}_1 & \boldsymbol{\beta}_1 & \boldsymbol{\gamma}_1 & \mathbf{t}_1 \\ 0 & 0 & 0 & 1 \end{bmatrix} \quad \mathbf{M}_2 = \begin{bmatrix} \boldsymbol{\alpha}_2 & \boldsymbol{\beta}_2 & \boldsymbol{\gamma}_2 & \mathbf{t}_2 \\ 0 & 0 & 0 & 1 \end{bmatrix} \tag{8}$$

Suppose that coordinate frame of Target 1 is reference, homogeneous expression of geometric relationship between two targets can be obtained as follow:

$$\mathbf{T}_t = \mathbf{M}_1^{-1}\mathbf{M}_2 \tag{9}$$

Let l be the distance between two origins, θ_x, θ_y and θ_z are the cosine values of the angles between three axis vector, we have

$$\begin{bmatrix} \theta_x & \theta_y & \theta_z & l \end{bmatrix} = \begin{bmatrix} \boldsymbol{\alpha}_1^T\boldsymbol{\alpha}_2 & \boldsymbol{\beta}_1^T\boldsymbol{\beta}_2 & \boldsymbol{\gamma}_1^T\boldsymbol{\gamma}_2 & \|\mathbf{t}_1 - \mathbf{t}_2\| \end{bmatrix} \tag{10}$$

As shown in Fig. 2, let $\boldsymbol{\Omega}_1$ and $\boldsymbol{\Omega}_2$ be the external matrices of Target 1 and 2 in Camera 1 and 2, respectively. $\mathbf{T}_c$ can be roughly obtained as follow:

$$\mathbf{T}_c = \boldsymbol{\Omega}_1\mathbf{T}_t\boldsymbol{\Omega}_2^{-1} \tag{11}$$

3.3   Accurate Estimation

In the previous we present the rough estimation of geometry relationships, which serve as initial values for accurate optimization in this subsection. Suppose that the

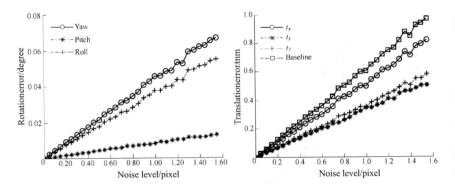

Fig. 2 Errors in global calibration with respect to the noisy data

apparatus is moved $N(N > 3)$ times in the calibration field, in each position we obtain the external matrices of Target 1 and 2 in Camera 1 and 2, $\mathbf{\Omega}_{1i}$ and $\mathbf{\Omega}_{2i}$, by decomposing the matrices as $\boldsymbol{\alpha}_{1i}, \boldsymbol{\beta}_{1i}, \boldsymbol{\gamma}_{1i}, \mathbf{t}_{1i}$ and $\boldsymbol{\alpha}_{2i}, \boldsymbol{\beta}_{2i}, \boldsymbol{\gamma}_{2i}, \mathbf{t}_{2i}$ referring to Eq. (1), $i = 1, 2...N$, objective functions contained $\mathbf{r}_{c21}$ can be established as follows:

$$
\begin{cases}
J = \sum_{i=1}^{N} \left\| g_{\alpha i} \right\|^2 + \sum_{i=1}^{N} \left\| g_{\beta i} \right\|^2 + \sum_{i=1}^{N} \left\| g_{\gamma i} \right\|^2 \\
g_{\alpha i} = \boldsymbol{\alpha}_{1i}^T \mathbf{r}_{c21} \boldsymbol{\alpha}_{2i} - \theta_x \\
g_{\beta i} = \boldsymbol{\beta}_{1i}^T \mathbf{r}_{c21} \boldsymbol{\beta}_{2i} - \theta_y \\
g_{\gamma i} = \boldsymbol{\gamma}_{1i}^T \mathbf{r}_{c21} \boldsymbol{\gamma}_{2i} - \theta_z
\end{cases}
\tag{12}
$$

From Eq. (6) we have

$$
\begin{cases}
\frac{\partial g_{\alpha i}}{\partial \boldsymbol{\psi}} = -\boldsymbol{\alpha}_{1i}^T (\mathbf{r}_{c21} \boldsymbol{\alpha}_{2i})^\times \\
\frac{\partial g_{\beta i}}{\partial \boldsymbol{\psi}} = -\boldsymbol{\beta}_{1i}^T (\mathbf{r}_{c21} \boldsymbol{\beta}_{2i})^\times \\
\frac{\partial g_{\gamma i}}{\partial \boldsymbol{\psi}} = -\boldsymbol{\gamma}_{1i}^T (\mathbf{r}_{c21} \boldsymbol{\gamma}_{2i})^\times
\end{cases}
\tag{13}
$$

Furthermore, $\partial g_{\alpha i}/\partial \theta_x = \partial g_{\beta i}/\partial \theta_y = \partial g_{\gamma i}/\partial \theta_z = -1$, Gauss-Newton iteration is used for nonlinear optimization from Eq. (12). The detailed steps are listed as follows:

Step 1. Get initial values of $\mathbf{r}_{c21}, \theta_x, \theta_y$ and θ_z from Sect. 3.2.
Step 2. Compute Jacobian matrices in Eq. (13) with current values.
Step 3. Obtain solutions of the following equations, $\Delta\boldsymbol{\psi}, \Delta\theta_x, \Delta\theta_y$ and $\Delta\theta_z$:

$$
\begin{aligned}
\frac{\partial g_{\alpha i}}{\partial \boldsymbol{\psi}} \Delta\boldsymbol{\psi} - \Delta\theta_x &= -g_{\alpha i} \\
\frac{\partial g_{\beta i}}{\partial \boldsymbol{\psi}} \Delta\boldsymbol{\psi} - \Delta\theta_y &= -g_{\beta i}, \quad i = 1, 2 \cdots N \\
\frac{\partial g_{\gamma i}}{\partial \boldsymbol{\psi}} \Delta\boldsymbol{\psi} - \Delta\theta_z &= -g_{\gamma i}
\end{aligned}
\tag{14}
$$

Step 4. Define $\Delta\phi = \|\Delta\boldsymbol{\psi}\|$, $\mathbf{k} = \Delta\boldsymbol{\psi}/\Delta\phi$, $\Delta\mathbf{R} = \mathbf{R}(\mathbf{k}, \Delta\phi)$ referring to Sect. 2, update $\theta_x \leftarrow \theta_x + \Delta\theta_x$, $\theta_y \leftarrow \theta_y + \Delta\theta_y$, $\theta_z \leftarrow \theta_z + \Delta\theta_z$ and $\mathbf{r}_{c21} \leftarrow (\Delta\mathbf{R})\mathbf{r}_{c21}$.
Step 5. Consider the small positive integrals ε_θ and ε_ϕ, if $\Delta\phi$ are less than ε_ϕ, $\Delta\theta_x$, $\Delta\theta_y$ and $\Delta\theta_z$ are less than ε_θ, break it, otherwise, return to Step 2.
Objective function contained $\mathbf{t}_{c21}$ can be established as follow:

$$
\begin{cases}
\kappa = \sum_{i=1}^{N} \left\| g_{ti} \right\|^2 \\
g_{ti} = \left\| \mathbf{t}_{c21} + \mathbf{r}_{c21} \mathbf{t}_{2i} - \mathbf{t}_{1i} \right\| - l
\end{cases}
\tag{15}
$$

From Eq. (15) we have

$$\frac{\partial g_{ti}}{\partial t_{c21}} = \frac{(t_{c21} + r_{c21}t_{2i} - t_{1i})^T}{\|t_{c21} + r_{c21}t_{2i} - t_{1i}\|} \quad \frac{\partial g_{ti}}{\partial l} = -1 \tag{16}$$

t_{c21} can be obtained as the same as r_{c21}, in particular, the linear equations in Gauss-Newton iteration is given by:

$$\frac{\partial g_{ti}}{\partial t_{c21}} \Delta t_{c21} - \Delta l = -g_{ti}, \; i = 1, 2 \cdots N \tag{17}$$

4 Experiment

4.1 Synthetic Data

In order to test the stability of the proposed method in the presence of noise, we carry out an experiment with simulation data. The translation between the cameras is [−1500 mm, 20 mm, −20 mm], the rotation is [−5°, 15°, −5°]. The pattern is 6 × 6 chequerboard, the cell size is 33 mm, the distance, $l = 1600$ mm. Two cameras have the properties as follows: imager resolution is 2592 × 1944, principal point coordinates are [1296, 972], effective focal lengths are [4000, 4000]. The calibrating points in images are brought in Gaussian noise with 0 mean and σ standard deviation. We carried out 200 trials for the level of noise changed from 0 pixels to 1.5 pixels, in each trial position is randomly changed 16 times. As shown in Fig. 2, the results list that the rotation error is less than 0.01° and the baseline error is less than 0.16 mm for $\sigma = 0.2$.

4.2 Real Data

An apparatus is manufactured base on the described principle. As shown in Fig. 3, the pattern of target is 6 × 6 chequerboard, cell size is designed to 33 mm, the length of lever is designed to 1800 mm, each leg is equipped with two stretching screw for adjusting the altitude of each lever end and maintaining the stability of the apparatus.

As shown in Fig. 4, we use a 3D four-wheel alignment system (Manufacturer: Shenzhen 3excel Tech Co., Ltd; Model: T50) to test the calibration method and 10 trials are carried out. The standard deviations of the rotation and translation are [0.0031°, 0.0025°, 0.0045°] and [0.190 mm, 0.121 mm, 0.396 mm], the means are [1.609°, 3.267°, −12.155°] and [−1930.353 mm, 38.490 mm, −23.767 mm]. Dispersion of the components in axis Z are larger than others, it is reasonable for the properties of monocular vision system in the primary analysis.

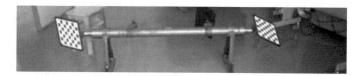

Fig. 3 Actual apparatus for global calibration

Fig. 4 Field calibration

Fig. 5 Standard frame

Figure 5 shows a standard frame for adjustment of the alignment system. The calibration accuracy is evaluated based on measurement results provided by the alignment system and the frame. Table 1. shows that the measurement errors of single toe-in in ± 3° and single camber in ± 5° are both less than 0.02°.

Table 1 Detection results of alignment angles, unit: degree

Alignment parameter	Standard	Mean	Deviation
Left toe-in	0.00	0.00	0.00
	1.50	1.49	−0.01
	3.00	3.00	0.00
	−1.50	−1.50	0.00
	−3.00	−3.00	0.00
Right toe-in	0.00	0.00	0.00
	1.50	1.49	−0.01
	3.00	2.99	−0.01
	−1.50	−1.48	+0.02
	−3.00	−2.99	+0.01
Left camber	0.00	0.00	0.00
	2.50	2.48	−0.02
	5.00	4.98	−0.02
	−2.50	−2.49	+0.01
	−5.00	−4.99	+0.01
Right camber	0.00	0.00	0.00
	2.50	2.50	0.00
	5.00	5.01	+0.01
	−2.50	−2.49	+0.01
	−5.00	−4.99	+0.01

5 Summary

A new method is proposed for global calibration of multi-camera measurement system from non-overlapping views based on constantness of the distance between original points and the angles between axis vectors of two targets. Simulated data shows that rotation error is less than 0.01° and the baseline error is less than 0.16 mm for noise level are 0.2 pixels. An associated apparatus is manufactured and a 3D four-wheel alignment system is used for carrying out real experiment, a standard frame is set up to detect the aligner after calibration, and measurement errors of single toe-in in ± 3° and single camber in ± 5° are both less than 0.02°.

Acknowledgements This work is partially supported by the Hunan Provincial Natural Science Foundation(No: 2015JJ5009), and the National Nature Science Foundation of China(No: 51405154, No: 51275169).

References

1. Muhlich, M., Aach, T.: High accuracy feature detection for camera calibration: a multi-steerable approach. In: Proceeding of the 29th DAGM Conference on Pattern Recognition, pp. 284–293. Springer-Verlag, Berlin, Heidelberg (2007)
2. de Franca, J.A., Stemmer, M.R.: A new robust algorithmic for multi-camera calibration with a 1D object under general motions without prior knowledge of any camera intrinsic parameter. Pattern Recogn. **45**, 3636–3647 (2012)
3. Olsen, B.D., Hoover, A.: Calibrating camera network using domino grid. Pattern Recogn. **34**, 1105–1117 (2001)
4. Shi, Y.Q., Sun, C.K., Wang, B. G., Wang, P., Duan, H.X.: A global calibration method of multi-vision sensors in the measurement of engine cylinder joint surface holes. In: International Conference on Materials Mechatronics and Automation, pp. 1182–1188. Melbourne, Australia (2011)
5. Wang, Y., Yuan, F., Jiang, H., Hu, Y.H.: Novel camera calibration based on cooperative target in attitude measurement. Optik **127**, 10457–10466 (2016)
6. Krüger, L., Wöhler, C.: Accurate chequerboard corner localisation for camera calibration. Pattern Recogn. Lett. **32**, 1428–1435 (2011)
7. Bouguet, J.T.: Camera calibration toolbox for matlab. MRL–Intel Corp. (2004)
8. Lu, H., Li, Y., Chen, M., Kim, H., Serikawa, S.: Brain intelligence: go beyond artificial intelligence. Mob. Netw. Appl. 1–10 (2017)
9. Lu, H., Li, Y., Mu, S., Wang, D., Kim, H., Serikawa, S.: Motor anomaly detection for unmanned aerial vehicles using reinforcement learning. IEEE Int. Things J. (2017) 10.1109/JIOT.2017.2737479
10. Serikawa, S., Lu, H.: Underwater image dehazing using joint trilateral filter. Comput. Electr. Eng. **40**(1), 41–50 (2014)

Leukemia Early Screening by Using NIR Spectroscopy and LAR-PLS Regression Model

Ying Qi, Zhenbing Liu, Xipeng Pan, Weidong Zhang, Shengke Yan, Borui Gan and Huihua Yang

Abstract In this paper, a regression analysis method based on the combination of Least Angle Regression (LAR) and Partial Least Squares (PLS) is proposed, which uses the non-invasive characteristics of near infrared spectroscopy (NIRS) to implement early screening of leukemia patients. First, the LAR method is used to eliminate collinearity between variables, second, PLS is employed to further build model for the wavelengths which are selected by the LAR. The result shows that this method needs less wavelength points and has more excellent performance in correlation coefficient and root mean square error, that are 0.9492 and 0.5917 respectively. The comparison experiments demonstrate that the LAR-PLS regression model has an advantage over principal component regression (PCR), the LAR-PCR regression model, successive projections algorithm (SPA) and elimination of uninformative variables (UVE) combined with PLS method in terms of predictive accuracy for screening leukemia patients.

Keywords Near-infrared spectroscopy · Least angle regression
Blood parameter · Early screening of leukemia · Partial least squares
Non-invasive

1 Introduction

Near-infrared spectroscopy has been used in the diagnosis of blood diseases as a non-invasive method [1, 2]. It has great potentials because it is non-invasive and the large number of in-depth NIR light can penetrate biological tissue [3]. Leukemia is

Y. Qi · Z. Liu (✉) · W. Zhang · S. Yan · B. Gan · H. Yang (✉)
School of Electronic Engineering and Automation, Guilin University of Electronic Technology, Guilin, China
e-mail: 3936924@qq.com

H. Yang
e-mail: 406611592@qq.com

X. Pan · H. Yang
School of Automation, Beijing University of Posts and Telecommunications, Beijing, China

© Springer International Publishing AG 2018
H. Lu and X. Xu (eds.), *Artificial Intelligence and Robotics*,
Studies in Computational Intelligence 752,
https://doi.org/10.1007/978-3-319-69877-9_21

193

a kind of cancer of white blood cells (WBC), which is mainly manifested in the bone marrow of abnormal proliferation of white blood cells or mitotic process [4]. However, some studies have shown that lactate dehydrogenase (LDH) is important risk factors for acute lymphoblastic leukemia (ALL).

This paper is to present a noninvasive method for screening leukemia based on the biomarker (LDH). The spectral data of leukemia and normal people are used to model, and the standard laboratory/reference data are utilized as response further to regression analysis. To achieve better variable selection in reducing variables of high dimensions, using the combination of LAR and PLS model has a better performance.

2   Related Work

Recently, a large body of work has emerged on the topic of the application about infrared spectroscopy in many fields. For example, a study has been done by Sahu et al. [5] the changes of spectral occurred in the white blood cells (WBC) of an adult acute myeloid leukemia (AML) patient. Ismail et al. [6] have proposed a non-invasive method in early screening of leukemia using NIRS. To eliminate the influence of instrumental noise and the co-linearity of variables these methods have been used commonly include: elimination of uninformative variables (UVE) [7], successive projections algorithm (SPA) [8] and interval PLS (iPLS) [9]. Kuang et al. [10] have presented a deep learning method for single infrared image stripe noise removal. Limmer et al. [11] have proposed a method for transferring the RGB color spectrum to near-infrared (NIR) images using deep multi-scale convolutional neural networks. Lu et al. also have developed a filtering deep convolutional network for marine organism classification [12–14]. An effective drug identification method by using deep belief network (DBN) with dropout mechanism (dropout-DBN) to model NIRS has been introduced by Yang et al. [15]. Monakhova et al. [16] have used NMR spectroscopy to distinguish pure sunflower lecithin from that blended with soy species, and to quantify the degree of such adulteration. Hence, a new non-invasive approach has been proposed which uses LAR combined with PLS prediction model in early screening of white blood patients, and achieved better results.

3   Methodology

3.1   Least Angle Regression Algorithm

The least angle regression determines that the coefficients of some variables are zero by constructing the first order penalty function. The linear regression model is as follows [17]:

$$\min S\left(\hat{\beta}\right) = \left\|\mathbf{y} - \hat{\boldsymbol{\mu}}\right\|^2 = \sum_{i=1}^{n} \left(y_i - \hat{\mu}_i\right)^2 = \sum_{i=1}^{n} \left(y_i - \sum_{j=1}^{p} x_{ij}\beta_j\right)^2 \quad (1)$$

$$subject\ to\ \sum_{j=1}^{p} |\beta_j| \leq t \quad (2)$$

where $(x_{i1}, x_{i2}, \ldots x_{ip})$ and y_i are the independent variables and response variables corresponding to the ith sample. The LAR algorithm minimizes y_i and the sum of the squared differences $S\left(\hat{\beta}\right)$ of the regression variables $\hat{\mu}_i$ by adjusting the values of β_j under Eq. (2).

3.2 The LAR-PCR Model

The proposed LAR-PCR model has a hierarchical structure, and it consists of two major layers. In the first layer the regression variable matrix β_j is obtained by Eqs. (1) and (2), the nonzero number β_k are replaced with 1 to update vector $\boldsymbol{\beta}$. $\mathbf{A}_i$ is vector which has l Column.

$$\mathbf{X}_i = \mathbf{A}_i\boldsymbol{\beta}, \quad i = 1, 2, \ldots, n \quad (3)$$

where $\mathbf{A}_i = [a_{i1}, a_{i2}, \ldots, a_{il}]$, $\beta = [\beta_1, \beta_2, \ldots, \beta_l]^T$
The second layer of the hierarchical structure is the regression.

$$y_j = \alpha_0 + \mathbf{X}_j\boldsymbol{\alpha} + \varepsilon_j, \varepsilon_j \sim N\left(0, \tau^2\right) \quad (4)$$

3.3 The LAR-PLS Model

The procedure of LAR and PLS proposed regression process is as follows:

(1) From the LAR linear regression model, the regression variable matrix β_j is obtained by Eqs. (1) and (2).
(2) Replace the nonzero number β_k with 1 to update vector $\boldsymbol{\beta}$. $\mathbf{X}_i$ is vector which has l Column. $\mathbf{q}_i$ is computed as:

$$\mathbf{q}_i = \mathbf{X}_i\boldsymbol{\beta}, i = 1, 2, \ldots, n \quad (5)$$

where $\mathbf{X}_i = [x_{i1}, x_{i2}, \ldots, x_{il}]$, $\boldsymbol{\beta} = [\beta_1, \beta_2, \ldots, \beta_l,]^T$

(3) $\mathbf{q}$ is updated spectral matrix. $\mathbf{Y}$ is actual response matrix. The matrices $\mathbf{T}$ (scores), $\mathbf{P}$ (loadings) and the vector $\mathbf{b}$ are computed sequentially in the next step.

$$\mathbf{Y} = \mathbf{TP}, \mathbf{q} = \mathbf{Tb} \tag{6}$$

(4) $\mathbf{q}$ is assumed to be a basis for $\mathbf{Y}$. Hence, we can approximate $\mathbf{Y}$ as $\mathbf{Y} = \mathbf{q}\mathbf{w}_{1,:}$, and calculate a best $\mathbf{w}_{1,:}$ (loading weights) in a linear least-squares fit. $\mathbf{t}_{:,1} = \mathbf{Y}/\mathbf{w}_{1,:}$, b_1 and $p_{1,:}$ are computed in linear least-squares fits.
(5) Both $\mathbf{Y}$ and $\mathbf{q}$ are projected onto $\mathbf{t}_{:,1}$ and the residuals calculated.

$$\mathbf{rq} = \mathbf{q} - \mathbf{t}_{:,1}b_1, \ \mathbf{Ry} = \mathbf{Y} - \mathbf{t}_{:,1}\mathbf{p}_{1,:} \tag{7}$$

(6) $\mathbf{Ry}$ replaces $\mathbf{Y}$ and the computation is continued at (4). The cycle (4) and (5) is repeated n times.

4 Experiments

MATLAB 2014a is employed as the software platform to implement comparative experiment. The compared results are showed to verify the validity of the proposed method.

4.1 Spectral and Laboratory/Reference Data

The data sets include the spectral data and laboratory/reference data. The spectral data consist of normal human's and leukemia's spectral data, and there are 60 groups. The wavelength range of the spectral data is between 900 and 2600 nm, which contains a total of 910 characteristic wavelength points. Laboratory/reference data is the value of blood parameter LDH which is actually measured in 60 groups respectively.

4.2 Spectral Data Pre-processing

The spectral data are pre-processed to enhance the quality of the acquired data by eliminate or minimize the effect of unwanted signal. In the beginning, using SPXY

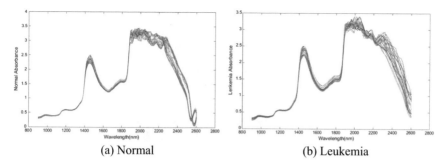

(a) Normal   (b) Leukemia

Fig. 1 Pre-processed spectral data: **a** Normal and **b** Leukemia

method the spectral data are divided into calibration set and prediction set, 40 samples and 20 samples respectively. Later on, Savitzky-Golay (SG) smoothing [18] is used to eliminate the interference of noise, the window size is 17. The pre-processed spectral data is shown in Fig. 1. It shows the results of the pre-processing of the normal spectral data in Fig. 1a. In Fig. 1b it shows the pre-processed results of leukemia patients.

4.3 Results and Discussion

The experimental results of four methods include Full-PLS, UVE-PLS, SPA-PLS and LAR-PLS are shown in Table 1. Besides, correlation coefficient of calibration (R_c^2), root mean square error of calibration (RMSEC), correlation coefficient of prediction R_p^2 and root mean square error of prediction (RMSEP) can be the valuation criteria. As can be seen from Table 1, LAR not only reduces the number of wavelength points, but also improves the prediction accuracy. Furthermore, R_p^2 of normal subjects increases from 0.8887 to 0.9492, and R_p^2 of leukemia patients increases from 0.8255 to 0.9024, which fully indicate that the wavelength extracted by LAR-PLS regression method can strongly eliminate collinearity between variables and remove the excess variables. The original spectrum of the effective information can be fully represented by only 24 wavelengths, which means this method can substantially reduce the dimension required for modeling.

Table 2 shows the results of the LAR-PCR model and LAR-PLS model evaluated on Blood parameter LDH for the normal and leukemia patients, it can be seen that the predictive performance of PLS model is better than PCR model under the similar conditions. In addition, the best effect is LAR-PLS model in which LAR combines with PLS model in the normal data, that R_p^2 and RMSEP can reach in 0.9492 and 15.0662, respectively. Likewise, the proposed method is proved to be valid.

Table 1 Models' performance by using different wavelength selection methods on both the calibration and prediction set of LDH

Sample/number	Method	Wavelength points	Components	Calibration set		Prediction set	
				R_c^2	RMSEC	R_p^2	RMSEP
Normal/30	Full-PLS	910	**4**	0.9440	17.6215	0.8887	22.2925
	UVE-PLS	550	16	**0.9999**	**0.5767**	0.9151	21.6998
	SPA-PLS	**6**	6	0.9332	18.5129	0.9251	20.3807
	LAR-PLS	24	12	0.9592	15.0398	**0.9492**	**15.0662**
Leukemia/30	Full-PLS	910	**5**	0.8377	145.4723	0.8255	171.892
	UVE-PLS	272	**5**	0.8811	120.0979	0.7970	192.2518
	SPA-PLS	**12**	7	0.8887	139.8732	0.8794	148.168
	LAR-PLS	24	15	**0.9075**	**108.7780**	**0.9024**	**129.8142**

Table 2 Models' performance by using LAR-PCR and LAR-PLS model on both the calibration and prediction set of LDH

Sample/number	Method	Wavelength points	Components	Calibration set		Prediction set	
				R_c^2	RMSEC	R_p^2	RMSEP
Normal/30	Full-PCR	910	4	0.9278	20.0025	0.8471	26.1310
	Full-PLS	910	4	0.9440	17.6215	0.8887	22.2925
	LAR-PCR	24	4	0.9119	22.1038	0.8343	27.2018
	LAR-PLS	24	12	**0.9592**	**15.0398**	**0.9492**	**15.0662**
Leukemia/30	Full-PCR	910	4	0.7305	221.5235	0.6089	217.7803
	Full-PLS	910	5	0.8377	171.892	0.8255	145.4723
	LAR-PCR	24	4	0.8278	161.6920	0.8053	201.6730
	LAR-PLS	24	15	**0.9075**	**108.7780**	**0.9024**	**129.8142**

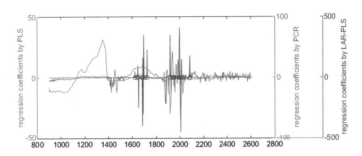

Fig. 2 Comparison of regression coefficients by using different models of LDH

In Fig. 2 the different performance of PCA, PLS and LAR-PLS for regression coefficients are showed by orange, blue and red lines, respectively. In addition, the red circle represents wavelength points selected by LAR-PLS model. It can be seen from Fig. 2 that the absorption rate relates to the blood parameter LDH in wave band of (649–651 nm) and (766–767 nm). The above wave bands have a greater effect on the blood parameter LDH. And these bands are selected as wavelength points which can achieve in LAR-PLS model. The regression coefficients of LAR-PLS model which are scattered in several points are obviously sharper than PCR and PLS. Therefore, LAR combined with PLS method is more likely to find optimal wavelength points and has better performance.

5   Conclusions

A method for the screening of leukemia in its early times via near infrared spectroscopy is investigated in this paper. A high degree of accurate results can be reached by the LAR-PLS method. UVE is poor in the screening of wavelengths that have no noise and no contribution, and the SPA cannot eliminate the wavelengths that are insensitive to the information to be measured. Comparing with the full spectrum PLS, this method is more likely to obtain the characteristic wavelengths that are mostly related with the dependent variable and explains the spectrum better. It can be inferred from the prediction results that the non-invasive method is proposed in leukemia screening. This method will partly focus on the biomarkers (LDH) by near infrared spectroscopy analytical technology. When the value of LDH increases, the risk of leukemia increases. The performance of the prediction model demonstrates NIR spectroscopy is a new non-invasive, simple and rapid technique for screening leukemia in its early times based on blood parameters.

Acknowledgements We are grateful for the financial support of the National Natural Science Foundation of China (No. 21365008, 61562013). We also thank Ismail R Faculty of Electrical Engineering for useful the spectral data and the laboratory/reference data.

References

1. Alam, M.K., Robinson, M.R.: Near-infrared noninvasive determination of pH in pulse mode. US, US 6542762 (2003)
2. Alam, M.K., Robinson, M.R.: Near-infrared noninvasive spectroscopic determination of pH. US, US 5792050 (1998)
3. Kraitl, J., Klinger, D., Fricke, D.M.: Non-invasive Measurement of Blood Components. Advancement in Sensing Technology, pp. 253–257. Springer, Berlin Heidelberg (2013)
4. Schultz, C.P., Liu, K., Johnston, J.B.: Study of chronic lymphocytic leukemia cells by FT-IR spectroscopy and cluster analysis. Leuk. Res. **20**(8), 649 (1996)
5. Sahu, R.K., Zelig, U., Huleihel, M.: Continuous monitoring of WBC (biochemistry) in an adult leukemia patient using advanced FTIR-spectroscopy. Leuk. Res. **30**(6), 687–693 (2006)
6. Ismail, R., Rahim, H.A., Rahim, I.M.A.: Near infrared spectroscopy (NIRS) applications in medical: non-invasive and invasive leukemia screening. J. 78(7-4) (2016)
7. Koshoubu, J., Iwata, T., Minami, S.: Elimination of the uninformative calibration sample subset in the modified UVE (Uninformative Variable Elimination)-PLS (Partial Least Squares) method. Anal. Sci. Intl. J. Jpn. Soc. Anal. Chem. **17**(2), 319 (2001)
8. Soares, S., Gomes, A.A.: The successive projections algorithm. Trends Anal. Chem. **42**(42), 84–98 (2013)
9. Li, Y., Mao, H., Shi, J.: Genetic algorithm interval partial least squares regression combined successive projections algorithm for variable selection in near-infrared quantitative analysis of pigment in cucumber leaves. Appl. Spectrosc. **64**(7), 786 (2010)
10. Kuang, X., Sui, X., Chen, Q.: Single infrared image stripe noise removal using deep convolutional networks. IEEE Photonics J. **9**(4), 1–13 (2017)
11. Limmer, M., Lensch, H.P.A.C. Infrared colorization using deep convolutional neural networks. In: IEEE International Conference on Machine Learning and Applications (2017)
12. Lu, H., Li, Y., Chen, M.: Brain intelligence: go beyond artificial intelligence (2017)
13. Lu, H., Li, B., Zhu, J.: Wound intensity correction and segmentation with convolutional neural networks. Concurrency Comput. Pract. Experience (2016)
14. Lu, H., Li, Y., Uemura, T.: FDCNet: filtering deep convolutional network for marine organism classification. J. Multimedia Tools Appl. 1–14 (2017)
15. Yang, H., Hu, B., Pan, X.: Deep belief networks based drug identification using near infrared spectroscopy. J. Innovative Opt. Health Sci. (2016)
16. Monakhova, Y.B., Diehl, B.W.K.: Quantitative analysis of sunflower lecithin adulteration with soy species by NMR spectroscopy and PLS regression. J. Am. Oil Chem. Soc. **93**(1), 27–36 (2016)
17. Shen, K., Bourgeat, P., Dowson, N.C.: Atlas selection strategy using least angle regression in multi-atlas segmentation propagation. In: IEEE International Symposium on Biomedical Imaging: From Nano To Macro, pp. 1746–1749. IEEE (2011)
18. Patchava, K.C, Alrezj, O., Benaissa, M.C.: Savitzky-Golay coupled with digital band pass filtering as a pre-processing technique in the quantitative analysis of glucose from near infrared spectra. In: 38th Annual International Conference of the IEEE Engineering in Medicine and Biology Society (EMBC), pp. 6210–6213 (2016)
19. Xu, X., He, L., Shimada, A.: Learning unified binary codes for cross-modal retrieval via latent semantic hashing. Neurocomputing **213**, 191–203 (2016)

Near Infrared Spectroscopy Drug Discrimination Method Based on Stacked Sparse Auto-Encoders Extreme Learning Machine

Weidong Zhang, Zhenbing Liu, Jinquan Hu, Xipeng Pan,
Baichao Hu, Ying Qi, Borui Gan, Lihui Yin, Changqin Hu
and Huihua Yang

Abstract This paper describes a method for drug discrimination with near infrared spectroscopy based on SSAE-ELM. ELM instead of the BP was introduced to fine-tuning SSAE, which can reduce the training time of SSAE and improve the practical application of the deep learning network. The work in the paper used near infrared diffuse reflectance spectroscopy to identify Aluminum-plastic packaging of cefixime tablets drugs from different manufacturers as examples to verify the proposed method. Specifically, we adopted SSAE-ELM to binary and multi-class classification discriminations with different sizes of drug dataset. Extensive experiments were conducted to compare the performances of the proposed method with ELM, BP, SVM and SWELM. The results indicate that the proposed method not only can obtain high discrimination accuracy with superior stability but also reduce the training time of SSAE in binary and multi-class classification. Therefore, the SSAE-ELM classifier can achieve an optimal and generalized solution for spectroscopy identification.

Keywords Near infrared spectroscopy (NIRS) · Stacked sparse auto-encoders extreme learning machine (SSAE-ELM) · Back propagation neural network (BP) Summation wavelet extreme learning machine (SWELM) · Drug classification

W. Zhang · Z. Liu (✉) · B. Hu · Y. Qi · B. Gan · H. Yang (✉)
School of Computer and Information Security, Guilin University
of Electronic Technology, Guilin 541004, China
e-mail: 3936924@qq.com

H. Yang
e-mail: 406611592@qq.com

J. Hu · X. Pan · H. Yang
School of Automation, Beijing University of Posts & Telecommunications,
Beijing 100876, China

L. Yin · C. Hu
National Institutes for Food and Drug Control, Beijing 100050, China

© Springer International Publishing AG 2018
H. Lu and X. Xu (eds.), *Artificial Intelligence and Robotics*,
Studies in Computational Intelligence 752,
https://doi.org/10.1007/978-3-319-69877-9_22

1 Introduction

Since NIRs contains abundant information of molecular groups of frequency doubling and frequency vibration. NIRs analysis technology can be used for fast and nondestructive detection of the drug [1].

With the development of pattern recognition technology, considerable NIRS methods have been widely used for the rapid and nondestructive quantitative and classification analysis. For instances, Deconinck et al. [2] constructed decision tree classifier model and obtained good classification accuracy from the spectroscopy of Viagra and Cialis drugs. SVM was adopted by Storme et al. [3] to build the classifier model to realize the drug discrimination of NIRS, whose accuracy was superior to the classification tree. SWELM was proposed by Liu et al. [4] to classify erythromycin ethylsuccinate NIRs of drugs produced by Xi'an Li Jun factory, and a new kernel function for signal feature extraction was built.

Deep learning is one of machine learning methods, and successfully used in multiple areas [5–8]. The deep learning models are especially suitable for processing high dimensional and nonlinear data since they have special deep network structure and nonlinear activation function. There are few papers focused on deep learning in the spectral classification and as our best knowledge, Yang et al. [9] employed sparse denoising auto-encoder (SDAE) to conduct discrimination of fake drugs. A SSAE is built by stacking additional unsupervised feature learning layers, and can be trained with greedy methods for each additional layer. ELM is a new single layer feed-forward neural network proposed by Huang et al. [10].

The accuracy is not high of small sample classification with limited extraction characteristics. In this paper, a method of drug identification of near infrared spectroscopy based on SSAE-ELM was proposed. To verify the validity of the algorithm, experiments were conducted based on diffuse reflection spectrum data of some drugs. Thus, we established the accuracy, stability and training time of SSAE-ELM, compared with the characteristics of ELM, SWELM, BP and SVM, and found the proposed method in this paper was effective.

2 Methods

2.1 Extreme Learning Machine

ELM can be regarded as much more generalized cluster of SLFNs whose hidden layer cannot be adjusted. ELM tends to reach both the minimum training error and the minimum norm of output weights. For N samples $(\mathbf{x}_k, \mathbf{t}_k)$, where $\mathbf{x}_k = [x_{k1}, x_{k2}, \ldots, x_{kn}]^T \in R^n$ and $\mathbf{t}_k = [t_{k1}, t_{k2}, \ldots, t_{km}]^T \in R^m$, standard ELM layer with $\hat{N}$ hidden nodes and activation function $g(x)$ are mathematically modeled as

$$\sum_{k=1}^{\hat{N}} \beta_k g_k (x_j) = \sum_{k=1}^{\hat{N}} \beta_k g(w_k \cdot x_j + b_k) = o_j, j = 1, 2, \cdots, N \qquad (1)$$

where $w_k = [w_{k1}, w_{k2}, \ldots, w_{kn}]$ is the weight vector connecting the kth hidden node and the input nodes, while $\beta_k = [\beta_{k1}, \beta_{k2}, \ldots, \beta_{km}]^T$ is the weight vector connecting the kth hidden node and the output nodes, and b_k is the threshold of the kth hidden node. The above N equations can be written compactly as

$$H\beta = T \qquad (2)$$

where $T = [t_1, \ldots, t_N]^T$ are the target labels of input data and $H = [h^T(x_1), \ldots, h^T(x_N)]^T$. To minimize the output error of ELM algorithm, ELM minimizes the training errors as well as the norm of the output weights according to:

$$\|H(\hat{w}_1, \ldots, \hat{w}_{\hat{N}}, \hat{b}_1, \ldots, \hat{b}_{\hat{N}})\hat{\beta} - T\| = \min_{w_k, b_k, \beta} \|H(w_1, \ldots, w_{\hat{N}}, b_1, \ldots, b_{\hat{N}})\beta - T\|$$

$$(3)$$

the above linear system is:

$$\hat{\beta} = H^{\dagger} T. \qquad (4)$$

where $H^{\dagger}$ is the Moore-Penrose generalized inverse of output matrix H.

2.2 Sparse Auto-Encoder

Sparse auto-encoder that is a feed-forward neural network with unsupervised learning using back propagation and batch gradient descent algorithm, attempts to learn an approximation to one identity function between the input and output data.

Suppose we have m fixed training examples $\{(x^{(1)}, y^{(1)}), (x^{(2)}, y^{(2)}), \ldots, (x^{(m)}, y^{(m)})\}$, the overall cost function of sparse auto-encoder can be defined by:

$$J_{sparse}(W, b) = \left[\frac{1}{m} \left(\frac{1}{2} \left\| h_{w,b}(x^{(i)}) - y^{(i)} \right\|^2 \right) \right] + \frac{\lambda}{2} \sum_{l=1}^{n_l - 1} \sum_{i=1}^{s_l} \sum_{j=1}^{s_l + 1} \left(w_{ji}^{(l)} \right)^2 + \varphi \sum_{j=1}^{s_l} KL(\rho \| \hat{\rho}_j)$$

$$(5)$$

where ρ is a sparsity parameter and $\hat{\rho}_j$ is the average activation of hidden unit j, Sparsity penalty term is used to approximate $\hat{\rho}_j$ to ρ which is a small value close to zero.

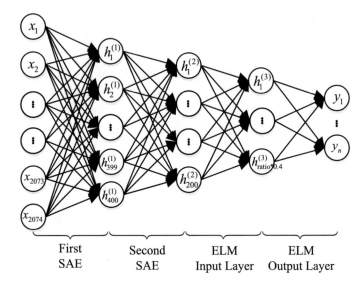

Fig. 1 The structure of SSAE-ELM model

2.3 SSAE-ELM Model

For the newly proposed SSAE-ELM, SSAE is employed to initialize the SSAE-ELM and learn the useful features from input drug NIRs data. We introduce ELM instead of the BP to fine tune SSAE. The training procedure of SSAE-ELM is shown in Fig. 1. The first three and last three layers of SSAE-ELM are SSAE and ELM, respectively, both of which share the third layer of SSAE-ELM. That is, the characteristics obtained from SSAE through the greedy access by the layers, are as the input for ELM. Finally SSAE-ELM is fine-tuned by the real label and the output of SSAE to ultimately achieve the ELM layer classification.

3 Experiments

3.1 Dataset and Experimental Setting

Experimental data were collected by the National Institutes for Food and Drug Control, including Aluminum-plastic packaging cefixime tablets from Hunan Fangsheng pharmaceutical factory and other three pharmaceutical factories. A complete spectrum had 2074 absorbance points and the numbers of the NIRS sample is 198.

DeepLearnToolbox was selected for experiment with the software platform MATLAB R2014a. The detailed process is as follows:

(1) Pre-processing: The spectroscopy data of cefixime tablets was pre-processed and normalized by max-min method.

(2) Pre-training: The first three layers of the SSAE-ELM implemented SSAE, where the SSAE structure was set as 2074-400-200. In the stage of SSAE training, iteration of each SSAE was 5 times and the learning rates of two layers were set to 0.05; the sparsity parameter and train were set to 0.1 and 0.4, and sigmoid function was used as activation function.

(3) ELM Fine-tuning: ELM with the structure 200-train * ratio-2/4 was adopted by the latter three layers of the SSAE-ELM. In the stage of ELM training, iteration of each SSAE was 1 time with the number of neurons train * 0.4 in the hidden layer and sigmoid was an activation function. The output of the SSAE is used as the input to the ELM in the SSAE-ELM, and SSAE-ELM is fine-tuned by the real label and the output of SSAE.

(4) Contrast experiments: The structure of ELM and SWELM were set to 2074-train * 0.4-2/4 and SVM selected radial basis function with Gaussian kernel parameters c = 1 and gamma = 0.3, while the two-layer BP structure was set to 2074-400-200-2/4 with the learning rate 0.01, sigmoid as activation function and iteration times of 50.

3.2 Binary Classification Experiment

Firstly, we tested the ability of SSAE-ELM model to predict identification of the genuine and fake drugs. Datasets of the collected 198 spectrum samples. 56 cefixime tablet spectral specimens of aluminum-plastic from Jiangsu Zhengda Qingjiang pharmaceutical factory as the positive set and 132 cefixime tablet spectral specimens of aluminum-plastic from other three pharmaceutical factories as the negative set. To further check the prediction performance of the algorithm in the datasets of different scale. By the random sampling, each dataset in Table 1 was configured and datasets were built for 10 times independently to evaluate their average performance.

Table 1 The size configuration of training sample set for binary-class discrimination

Number of training sample in total	Number of positive sample	Number of negative sample
60	20	40
80	25	55
100	30	70
120	35	85
140	40	100
160	45	115

Table 2 The binary-classification accuracy on different ratios of training samples (unit: %)

Train/Test data	ELM	SVM	SWELM	BP	SSAE-ELM
60/156	93.84	**98.97**	96.02	97.43	97.31
80/136	94.41	97.64	91.32	95.88	**100**
100/116	93.44	98.67	94.13	98.27	**99.31**
120/96	91.87	99.78	93.54	**100**	99.79
140/76	90	97.89	96.05	**100**	99.47
160/56	94.64	**100**	95	98.57	**100**

In Table 2, the results indicate that the accuracy of ELM fine-tuning SSAE was higher than that of ELM and SWELM, especially for the small samples SSAE-ELM had the best performance. This suggests that SSAE-ELM could effectively improve the classification accuracy of the model. As shown in Table 2, SVM and r BP had similar high accuracy, demonstrating that their complex nonlinear modeling capability could be applied to the binary classification problem. However, the ELM and SWELM did not own a nonlinear modeling capability, therefore had lower prediction accuracy.

The same procedures were used for algorithm stability and training time, as shown in Fig. 2a, b. The SSAE-ELM had enhanced stability during the whole training process, which was superior than ELM and SWELM. With the increase of training samples, the stability of SSAE-ELM was generally higher than that the other four algorithms. Since BP was replaced by ELM to fine-tune SSAE, which can reduce the training time of SSAE, therefore the SSAE-ELM possessed improved advantages than BP. However, as extract spectral features in pre-training were not essential for ELM, SWELM and SVM, which spent less time.

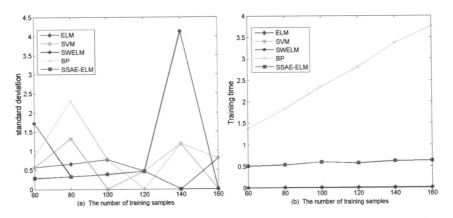

Fig. 2 The STD of accuracy of different binary classification models and training time of different binary classifier on different ratios of training samples

Table 3 The size configuration of training samples for multi-class discrimination

Training sample	Class 1 (63)	Class 2 (54)	Class 3 (51)	Class 4 (48)
60	20	15	13	12
80	25	20	18	17
100	30	25	23	22
120	35	30	28	27
140	40	35	33	32
160	45	40	38	37

3.3 Multi-class Classification Experiment

Secondly, we took multi-stage classification of drug identification as an example to test the prediction ability of SSAE-ELM model. The spectrum samples were divided into the following as four classes in Table 3. The spectral samples of aluminum-plastic packaging, in Hunan Fangsheng Pharmaceutical Co., Ltd., were taken as the first class. Then we let the spectral samples of the aluminum-plastic packaging as the second, third, and fourth class, from the other manufacturers.

Similar to the experiment of binary classification, the samples were randomly selected, and the training and test sets were built according to the sizes of datasets and the amount of data in different categories (see Table 3). By the random sampling, each dataset in Table 3 was configured and datasets were built for 10 times independently to evaluate their average performance.

In the case of classification accuracy, the average prediction accuracy was shown in Table 4, and the results implied that SSAE-ELM was advanced than that of ELM, SWELM and BP, especially for the smaller the sample. Meanwhile, the average accuracy rate of SSAE-ELM and ELM, SWELM in the multi-classification problem (Table 4) and the classification problem (Table 2) were further compared, and the advantages of SSAE-ELM in the multi-classification were more significant due to fewer samples of each category.

The same procedures were used for algorithm stability and training time, as shown in Fig. 3a, b. The results illustrate that the SSAE ELM and SVM can still maintain the high stability, which is generally greater than that of ELM, SWELM and BP. While the training time of ELM, SWELM and SVM was still rapid.

Table 4 The multi-class classification accuracy on different ratios of training samples

Train/Test data	ELM	SVM	SWELM	BP	SSAE-ELM
60/156	92.62	98.71	89.61	95.19	**100**
80/136	90.66	98.52	93.6	94.26	**99.33**
100/116	92.93	98.36	89.48	98.01	**100**
120/96	89.06	98.95	88.02	97.91	**99.89**
140/76	88.94	98.68	86.05	97.36	**100**
160/56	96.42	**100**	91.42	99.46	**100**

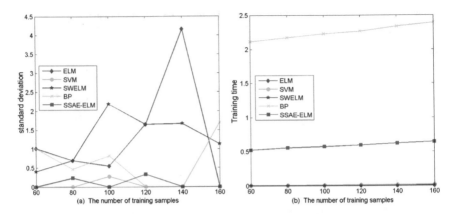

Fig. 3 The STD of accuracy of different multi-class classification models and training time of different multi-class classifier on different ratios of training samples

4 Conclusions

According to the study of binary and multi-class classification on near infrared diffuse reflectance spectroscopy of drugs, SSAE-ELM not only has better classification accuracy and stability but also reduce the training time of SSAE, and the advantages are more substantial with the more sufficient data.

Acknowledgements We are grateful for the financial support of the National Natural Science Foundation of China (No. 21365008, 61562013).

References

1. Risoluti, R., Materazzi, S., Gregori, A., et al.: Early detection of emerging street drugs by near infrared spectroscopy and chemometrics. J. Talanta **153**, 407 (2016)
2. Deconinck, E., Sacré, P., Coomans, D., et al.: Classification trees based on infrared spectroscopic data to discriminate between genuine and counterfeit medicines J. J. Pharm. Biomed. Anal. **57**, 68–75 (2012)
3. Storme, P.I., Rebiere, H.: Challenging near infrared spectroscopy discriminating ability for counterfeit pharmaceuticals detection. Anal. Chim. Acta **658**(2), 163–174 (2010)
4. Liu, Z.B., Jiang, S.J., Yang, H.H., et al.: Drug discrimination by near infrared spectroscopy based on summation wavelet extreme learning machine. Spectrosc. Spectr. Anal. **34**(10), 2815 (2014)
5. Xu, X., He, L., Shimada, A., et al.: Learning unified binary codes for cross-modal retrieval via latent semantic hashing. Neurocomputing **213**, 191–203 (2016)
6. Lu, H., Li, Y., Uemura, T., et al.: FDCNet: filtering deep convolutional network for marine organism classification. Multimedia Tools Appl. 1–14 (2017)
7. Lu, H., Li, B., Zhu, J., et al.: Wound intensity correction and segmentation with convolutional neural networks. Concurr. Comput. Pract. Exp. (2016)
8. Lu, H., Li, Y., Chen, M.: Brain Intelligence: Go Beyond Artificial Intelligence. J. (2017)

9. Yang, H.H., Luo, Z.C., et al.: Sparse denoising autoencoder application in identification of counterfeit pharmaceutical. J. Spectrosc. Spectr. Anal. **36**, 2274–2779 (2016)
10. Huang, G.B., Zhu, Q.Y., Siew, C.K.: Extreme learning machine: Theory and applications. J. Neurocomput. **70**(1–3), 489–501 (2006)

A Concise Conversion Model
for Improving the RDF Expression
of ConceptNet Knowledge Base

Hua Chen, Antoine Trouve, Kazuaki J. Murakami and Akira Fukuda

Abstract With the explosive growth of information on the Web, Semantic Web and related technologies such as linked data and commonsense knowledge bases, have been introduced. ConceptNet is a commonsense knowledge base, which is available for public use in CSV and JSON format; it provides a semantic graph that describes general human knowledge and how it is expressed in natural language. Recently, an RDF presentation of ConceptNet called ConceptRDF has been proposed for better use in different fields; however, it has some problems (e.g., information of concepts is sometimes misexpressed) caused by the improper conversion model. In this paper, we propose a concise conversion model to improve the RDF expression of ConceptNet. We convert the ConceptNet into RDF format and perform some experiments with the conversion results. The experimental results show that our conversion model can fully express the information of ConceptNet, which is suitable for developing many intelligent applications.

Keywords ConceptNet · RDF · ConceptRDF · AI · Knowledge base

H. Chen (✉) · K.J. Murakami · A. Fukuda
Graduate School and Faculty of ISEE, Kyushu University, Fukuoka, Japan
e-mail: hua.chen@socait.kyushu-u.ac.jp; chenhua5752@hotmail.com

K.J. Murakami
e-mail: murakami@ait.kyushu-u.ac.jp

A. Fukuda
e-mail: fukuda@ait.kyushu-u.ac.jp

A. Trouve · K.J. Murakami
Team AIBOD Co., Ltd., Fukuoka, Japan
e-mail: trouve@aibod.com

© Springer International Publishing AG 2018
H. Lu and X. Xu (eds.), *Artificial Intelligence and Robotics*,
Studies in Computational Intelligence 752,
https://doi.org/10.1007/978-3-319-69877-9_23

213

1   Introduction

The World Wide Web is an information space where documents and other web resources are interlinked by hypertext links, and can be accessed via the Internet. The goal of the World Wide Web is to create a most appropriate presentation of information to facilitate the communication between humans and computers.

In the past three decades, much progress has been made about the World Wide Web. Web 1.0 is the first generation of web and could be considered a "read-only" web. It focuses on homepages to broadcast information to users and lacks of interactions between information creators and users. Web 2.0 was defined in 2004 as a "read-write" web. In the era of Web 2.0, people often consume as well as contribute information through different APIs, blogs or sites such as Facebook, Flickr and YouTube. Web 2.0 also has some limitations: it only supports keyword-based search; information in each Website is separate but not shared; computers cannot understand the information automatically. Web 3.0 desires to decrease human's tasks and decisions and leave them to machines by providing machine-readable contents on the web [1]. Web 3.0 is also known as Semantic Web, in which the information is expressed in RDF (Resource Description Framework) [2] format that can be understood and reasoned by computers.

Besides RDF, RDFS [3], computational ontology and artificial intelligence (AI) technologies are also used in Semantic Web to provide specific domain knowledge base and support intelligent inference in different fields. It is obvious that a rich knowledge base constitutes an essential element for the development of intelligent applications.

ConceptNet [4] is a large commonsense knowledge base, which provides a large semantic graph that describes general human knowledge and how it is expressed in natural language. It can be used in many different fields, such as solving the *Symbol Grounding Problem* [5] in AI applications, developing Question Answering systems [6], image tagging [7] and semantic retrieval [8]. ConceptNet is available for public use in CSV and JSON [9] format and through its Web site and API. However, in some cases, people may need to convert ConceptNet into RDF format: this is still a challenge due to the complexity of its edges. In [10], Grassi et al. proposed to encode ConceptNet in RDF to make it directly available for intelligent applications, they also discussed the feasibility and benefits of this conversion. In [6, 11], Najmi et al. designed an upper ontology and proposed ConceptRDF: an RDF representation of ConceptNet knowledge base. They described the details of the conversion process and converted the whole dataset of ConceptNet into RDF expressions. ConceptRDF consists of 50 RDF files, which are available for public use.[1] A partial view of the ConcetptRDF is as follows (omitting prefix definitions).

---

[1]Refer to http://score.cs.wayne.edu/result/.

```
<http://conceptnet5.media.mit.edu/web/c/en/quantity>
        COnto:hasSource    "verbosity", "wordnet";
        COnto:hasPOS       "n", "a";
        COnto:hasSense     "present_in_great_quantity";
        COnto:hasAttribute "abundant".
```

In examining the sample RDF triples, we find that the information is sometimes misexpressed. For example, "present in great quantity" should be the sense of "abundant" but not "quantity" (will be presented later in Sect. 2). In this context, we need to find a way to improve the RDF expression of ConceptNet.

In this paper, we propose a concise conversion model to improve the RDF expression of ConceptNet. We introduce the conversion process and illustrate some use cases to show how to use our conversion results for developing intelligent applications (e.g., Question Answering systems). The experiment results show that our conversion model can fully express the information of ConceptNet, which is suitable for developing many intelligent applications.

The rest of this paper is organized as follows. Section 2 briefly introduces some commonsense knowledge bases and the structure of ConceptNet. In Sect. 3, we describe our conversion model and convert the ConceptNet into RDF format. In Sect. 4, we illustrate some use cases to show the merits of our conversion. Section 5 gives the conclusion and future work.

2 Background

Commonsense Knowledge Bases. Several approaches have been proposed to develop large knowledge bases. Cyc [12] is an oldest AI project for gathering information from everyday commonsense knowledge. Parts of the project are released as OpenCyc,[2] which provides an API, SPARQL [13] endpoint, and data dump under an open source license. Freebase [14] is a scalable tuple database used to structure general human knowledge. DBpedia [15] is a community effort to extract structured information from Wikipedia, it can be either imported into third party applications or can be accessed online using a variety of user interfaces. YAGO [16] is another project that extracts information from Wikipedia, WordNet (a knowledge base of English words) [17] and other resources; it can be queried through various browsers or through a SPARQL endpoint.

The knowledge bases mentioned above are almost focused on gathering "facts" in different fields, similar to encyclopedias. ConceptNet is another commonsense knowledge base, its scope includes words and common phrases in any written human language. These words and phrases are related through an open domain of

---

[2]As of 2017, OpenCyc is no longer available, details can refer to http://www.opencyc.org.

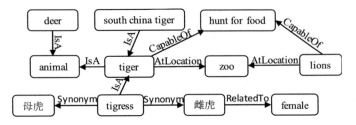

Fig. 1 Concepts and their relations from ConceptNet

```
/a/[/r/Attribute/,/c/en/quantity/n/an_adequate_or_large_amount/,/c/en/abund
ant/a/present_in_great_quantity/]    /r/Attribute
/c/en/quantity/n/an_adequate_or_large_amount
/c/en/abundant/a/present_in_great_quantity        /ctx/all 1.5849625007211563
/s/wordnet/3.0 /e/b0d2bb237b02a51bea5eb69c27acf9582da8dc91 /d/wordnet/3.0
                      (A) A relation in CSV format
```

```
{"end": "/c/en/abundant/a/present_in_great_quantity", "surfaceStart":
null, "weight": 1.5849625007211563, "id":
"/e/b0d2bb237b02a51bea5eb69c27acf9582da8dc91", "rel": "/r/Attribute",
"surfaceEnd": null, "surfaceText": null, "sources": ["/s/wordnet/3.0"],
"context": "/ctx/all", "dataset": "/d/wordnet/3.0", "source_uri":
"/s/wordnet/3.0", "features":
["/c/en/quantity/n/an_adequate_or_large_amount /r/Attribute -",
"/c/en/quantity/n/an_adequate_or_large_amount -
/c/en/abundant/a/present_in_great_quantity", "- /r/Attribute
/c/en/abundant/a/present_in_great_quantity"], "uri":
"/a/[/r/Attribute/,/c/en/quantity/n/an_adequate_or_large_amount/,/c/en/abu
ndant/a/present_in_great_quantity/]", "start":
"/c/en/quantity/n/an_adequate_or_large_amount", "license": "/l/CC/By-SA"}
                      (B) A relation in JSON format
```

Fig. 2 Same information presented in a line with two formats

predicates, describing not just how words are related by their lexical definitions, but also how they are related through common knowledge [18]. Figure 1 shows a cluster of related concepts and the ConceptNet assertions that connect them.

The Structure of ConceptNet. ConceptNet can be seen as a graph with a set of nodes and edges: the nodes represent concepts and the edges represent the relations between concepts. There are two formats available: one is CSV (as shown in Fig. 2a); the other is JSON, which containing the same information as CSV but in a better human-readable way (as shown in Fig. 2b).

We use the CSV format to describe the structure. For each line in CSV, it starts with an assertion tag "/a/"; other tags and their meanings are as follows:

- /r/: the relation between concepts;
- /c/: language-independent concept with a term or a phrase;
- /ctx/: the context of the relation;
- /d/: marks where the relation has been extracted from;

- /s/: presents the source of the information;
- /e/: shows the edge ID (each edge can be identified by an edge ID);
- /and/: marks the conjunctions of sources.
- /or/: marks the disjunctions of sources.

Besides the tags mentioned above, there is a number following the "/ctx/all", it represents the weight of an edge.

We also notice that each concept URI (Uniform Resource Identifier) consists of up to 4 parts; we separate the concept URI with "/" (omitting tag "/c/") and then get several smaller parts: (1) language mark (e.g. *en* for *English*, *ja* for *Japanese*); (2) its normalized text; (3) its POS (part-of-speech) tags (e.g., *n* for *noun*, *v* for *verb*); (4) the sense of the concept (available for the information extracted from WordNet).

With the descriptions above, we know the meaning of the information shown in Fig. 2: "quantity (written in English) is a noun, has the meaning of 'an adequate or large amount', is attribute of abundant, which is an adjective with the meaning of 'present in great quantity'; the context of the edge is 'all'; the weight of the edge is 1.5849625007211563; the source and the edge are extracted from WordNet version 3.0; its edge ID is b0d2bb237b02a51bea5eb69c27acf9582da8dc91".

3 Proposed Approach

3.1 Proposed Conversion Model

We use the CSV files to generate RDF files; each line in CSV represents a triple of subject, object and predicate. Generally, the subject and the object are concepts; the predicate is the relation between concepts. They can be converted into RDF expressions with definitions as follows.

Definition 1 For each concept, it consists of up to 4 parts: *language_mark, normalized_text, POS_tag* and *concept_sense;* they can be expressed with RDF triples as follows:

{(ConceptURI, rdfs:label,[3] *normalized_text@language_mark);*
(ConceptURI, hasPOS, POS_tag); (ConceptURI, hasSense, concept_sense)}.

Definition 2 For each relation, it has many properties, they can be expressed with RDF triples as follows (now, relation names are all expressed in English):

{(EdgeID,   rdfs:label,   Relation_name@en);   (EdgeID,   hasContex,
Contex_value);
(EdgeID, hasDataset, Dataset_value); (EdgeID, hasSource, Source_value);
(EdgeID, hasWeight, Weight_value)}.

---

[3]A predefined property, refer to http://www.w3.org/TR/rdf-schema/.

```
    </c/en/quantity/n/an_adequate_or_large_amount>
            RDFS:label"    quantity"@en;
            COnto:hasPOS"  n";
            COnto:hasSence"a n_adequate_or_large_amount".
    </c/en/abundant/a/present_in_great_quantity>
            RDFS:label"a    bundant"@en;
            COnto:hasPOS   "a";
            COnto:hasSence" present_in_great_quantity".
    <b0d2bb237b02a51bea5eb69c27acf9582da8dc91>
            RDFS:label"Attri     bute"@en ;
            COnto:hasContext "all";
            COnto:hasDataset"w ordnet";
            COnto:hasSource"w  ordnet";
            COnto:hasWeight  "1.5849625007211563";
            COnto:hasSubject </c/en/quantity/n/an_adequate_or_large_amount>;
            COnto:hasObject  </c/en/abundant/a/present_in_great_quantity>.
```

Fig. 3 RDF expressions of concepts and their relation

Definition 3 For each relation between two concepts, they can be expressed with the following two RDF triples:

{(EdgeID, hasSubject, sub_Concept); (EdgeID, hasObject, obj_Concept)}.

With these definitions, the information presented in Fig. 2 can be expressed with RDF triples, as shown in Fig. 3.

3.2 Conversion Process

The converting process is similar to the work in [6]. We download the CSV files from the public website[4] and read them line by line to extract information items (e.g., concept, relation, edgeID, etc.). Then, we model the information items with our conversion model. Finally, we generate a set of files with RDF expressions of the ConceptNet.

4 Use Cases

In this section, we illustrate some use cases to show how to use our conversion results of ConceptNet for developing intelligent applications.

---

[4]http://conceptnet5.media.mit.edu/downloads/current/.

4.1 Question Answering

Our conversion results can be used to answer questions about concepts. Table 1 shows the questions and their expressions of triples. The questions can be translated into SPARQL queries based on our definitions introduced in Sect. 3. For example, question No.1 can be translated into a SPARQL query as follows.

```
SELECT ?Sub WHERE {
  ?Sub_Concept  rdfs:label        ?Sub.
  ?Obj_Concept  rdfs:label        "smell_scent"@en.
  ?edgeId       rdfs:label        "CapableOf"@en.
  ?edgeId       COnto:hasSubject  ?Sub_Concept.
  ?edgeId       COnto:hasObject   ?Obj_Concept.}
```

With the SPARQL queries, the conversion results of ConceptNet can be queried to find answers. In our experiments, the answers are as follows:

- *No.1: human; dog.*
- *No.2: predator; animal; lion; person; tiger.*
- *No.3: n.*
- *No.4: an_adequate_or_large_amount.*

We also use ConceptRDF for Question Answering with the same questions. For question No. 1 and No. 2, the answers are the same; but for question No. 3 and No. 4, the answers are incorrect. These results show that our approach has improved the RDF expression of ConceptNet and solved the problem mentioned in Sect. 1.

Analysis. Our conversion model focuses on the expressions of both *concepts* and *relations*; however, the ConceptRDF does not deal with *relations* carefully, the information items of *relations* are incorrectly linked to the *concepts*.

Table 1 Questions and their expression of triples

No.	Questions	Expressed as triples
1	What can smell scent?	<?Sub, CapableOf, smell_scent>
2	What can hunt for food?	<?Sub, CapableOf, hunt_for_food>
3	What is the POS tags of quantity?	<quantity, hasPOS, ?tag>
4	What is the sense of quantity?	<quantity, hasSense, ?sense>

4.2 Multilingual Information Retrieval

Our conversion results can also be used for multilingual information retrieval. For example, "animal" and "动物" (in Chinese) have the same meaning, we can find the relations between them with a SPARQL query as follows:

```
SELECT ?relation  WHERE {
    ?subConcept rdfs:label      "动物"@zh.
    ?edgeId     rdfs:label      ?relation.
    ?edgeId     COnto:hasSubject ?subConcept.
    ?edgeId     COnto:hasObject  ?ObjConcept.
    ?ObjConcept rdfs:label      "animal"@en.}
```

With this SPARQL query, it returns "relatedTo" as a result. We check the information on the Web site of ConceptNet and find that "synonym" should also be a result but it cannot be found. We analyze the dataset of ConceptNet (CSV files) and find that some information (e.g., *synonyms* from WordNet) is not included in the dataset. Thus, we may need to extend ConceptNet with other resources (e.g., WordNet) when developing applications for multilingual information retrieval.

5 Conclusion and Future Work

In this paper, we proposed a concise conversion model to improve the RDF expression of ConceptNet. We also illustrated some use cases with the conversion results. The experiment results show that our conversion model can fully express the information of ConceptNet knowledge base, which is suitable for use in many intelligent applications.

In the future, we plan to improve our work in two aspects: (1) improve the upper ontology introduced in [11] to allow complex inferences in intelligent applications; (2) use our RDF expressions of conceptNet to integrate some other resources (e.g. images, videos) for building complex intelligent applications.

References

1. Aghaei, S., Nematbakhsh, M.A., Farsani, H.K.: Evolution of the World Wide Web: from Web 1.0 to Web 4.0. Int. J. Web & Semant. Technol. **3**(1), 1–3 (2012)
2. Manola, F., Miller, E., McBride, B.: Resource description framework (RDF) primer. W3C Recomm. (2004)
3. Brickley, D., Guha, R.V.: RDF Vocabulary description language 1.0: RDF schema. W3C Recomm. (2004). https://www.w3.org/TR/rdf-schema

4. Liu, H., Singh, P.: ConceptNet—a practical commonsense reasoning tool-kit. BT Technol. J. **22**(4), 211–226 (2004)
5. Lu, H., Li, Y., Chen, M., Kim, H., Serikawa, S.: Brain Intelligence: Go Beyond Artificial Intelligence. arXiv preprint (2017)
6. Najmi, E., Malik, Z., Hashmi, K., Rezgui, A.: ConceptRDF: An RDF presentation of ConceptNet knowledge base. In: 7th International Conference on Information and Communication Systems (ICICS), pp. 145–150. IEEE (2016)
7. Xu, X., He, L., Lu, H., Shimada, A., Taniguchi, R.I.: Non-linear matrix completion for social image tagging. IEEE Access **5**, 6688–6696 (2017)
8. Xu, X., He, L., Shimada, A., Taniguchi, R.I., Lu, H.: Learning unified binary codes for cross-modal retrieval via latent semantic hashing. Neurocomputing **213**, 191–203 (2016)
9. Crockford, D.: The Application/Json Media Type for Javascript Object Notation (json) (2006)
10. Grassi, M., Piazza, F.: Towards an RDF encoding of ConceptNet. In: International Symposium on Neural Networks, pp. 558–565. Springer, Berlin (2011)
11. Najmi, E., Hashmi, K., Malik, Z., Rezgui, A., Khanz, H.U.: ConceptOnto: an upper ontology based on conceptnet. In: 11th International Conference on Computer Systems and Applications (AICCSA), pp. 366–372. IEEE (2014)
12. Lenat, D.B.: CYC: A large-scale investment in knowledge infrastructure. Commun. ACM **38** (11), 33–38 (1995)
13. Prud, E., Seaborne, A.: SPARQL Query Language for RDF (2006)
14. Bollacker, K., Evans, C., Paritosh, P., Sturge, T., Taylor, J.: Freebase: A collaboratively created graph database for structuring human knowledge. In: The 2008 ACM SIGMOD International Conference on Management of Data, pp. 1247–1250. ACM (2008)
15. Auer, S., Bizer, C., Kobilarov, G., Lehmann, J., Cyganiak, R., Ives, Z.: DBpedia: A Nucleus for a Web of Open Data. The Semantic web, pp. 722–735 (2007)
16. Suchanek, F.M., Kasneci, G., Weikum, G.: Yago: A core of semantic knowledge. In: 16th International Conference on World Wide Web, pp. 697–706. ACM Press, USA (2007)
17. Miller, G.A.: WordNet: a Lexical database for English. Commun. ACM **38**(11), 39–41 (1995)
18. Speer, R., Havasi, C.: Representing general relational knowledge in ConceptNet 5. In: LREC, pp. 3679–3686 (2012)

Current Trends and Prospects of Underwater Image Processing

Jinjing Ji, Yujie Li and Yun Li

Abstract In view of the consumption and distribution of resources in the world, the necessity of deep-sea mining video image processing is illustrated. Based on a large number of relevant literatures, this paper summarizes the research status of image processing methods for deep-sea mining observation video system, introduces the advantages of the improved median filter algorithm, the improved dark channel prior algorithm and the improved nonlocal mean denoising method. Some problems of the original methods are analyzed. These aim to provide reference for the optimization and improvement of image processing methods for deep-sea mining video observation.

Keywords Video image · Image processing · Deep-sea mining Improved algorithm

1 Introduction

As the continuous exploitation of land resources, the world is gradually showing up a scarcity of resources, when people are seeking new energy, they also begin to research for ocean accounted for 71% of the earth [1–10]. In marine mineral resources, polymetallic nodules have a high exploitation value and economic benefits. Polymetallic nodules are rich in manganese, nickel, copper, cobalt and other metal elements, widely distributed in the seabed whose depth is 4000–6000 m of the Pacific Ocean, the Atlantic Ocean and the Indian Ocean, the total quantities of marine polymetallic nodules are about 3 million tons [11]. Until April 2016, the International Seabed Authority (ISA) has approved 27 applications for international seabed mining, including China, France, Japan, Russia, the United Kingdom, Germany, South Korea, India and other countries. The requirements for deep-sea

J. Ji · Y. Li (✉) · Y. Li
Yangzhou University, Yangzhou, China
e-mail: liyujie@yzu.edu.cn; yzyjli@gmail.com

© Springer International Publishing AG 2018
H. Lu and X. Xu (eds.), *Artificial Intelligence and Robotics*,
Studies in Computational Intelligence 752,
https://doi.org/10.1007/978-3-319-69877-9_24

223

mining technology will be one of the important bases for national development in the future, and countries will continue to strengthen deep-sea research.

At present, the investigation methods for deep-sea polymetallic nodules are as follows: multi-beam, heavy magnetism, earthquakes, submarine imaging and trawl, etc. As a new type of geological survey, submarine camera can be the more intuitive observation of the seabed topography, which has an important role in seabed survey, seabed mineral resources census [12–15]. However, light attenuation, absorption and scattering caused by seawater on the ocean, the presence of suspended solids in the ocean, will all lead difficulties of getting the underwater images. In short, underwater images are usually subject to uneven illumination, low contrast and noise, which are not conducive to the subsequent analysis and processing. It is the key of deep-sea mining. In order to further improve the imaging results, it is necessary to use digital image processing technologies for subsequent processing of underwater imaging.

2 Related Works

Because of its special imaging background, it is essential to do specialized algorithm research in underwater images. At present, the main techniques for underwater image processing include image restoration technology [16, 17] and image enhancement technology. Among them, the purpose of the image restoration technology is utilizing the degradation model and the restoration of the original imaging model. This type of method needs to take a variety of parameters into account, such as the support of the absorption coefficient of water and scattering coefficient. The applicability is relatively poor, to achieve more difficult. Another method is image enhancement technology, mainly using subjective qualitative to determine the quality of the image, the use of filters and other technologies to image processing more clearly, does not rely on the physical model of underwater imaging, usually more than image restoration technology For quick and easy.

3 Types and Characteristics of Improved Algorithms

Some classical algorithms such as nonlocal mean denoising methods, dark channel prior algorithm classical median filtering algorithms play an important role in underwater image processing. But with the development of the times, the increase in demand for deep sea mining, image processing also has a greater demand. Although these algorithms can solve the problem of image blur, filter denoising, color reproduction, but in general the processing is not accurate enough, or will lead to some errors in the processing of images, causing damage to mining. So the improvement in the algorithm has been the focus of deep-sea mining video observation system. At present, the improved algorithm mainly has improved nonlocal mean denoising method, improved dark primary a priori algorithm, and improved median filtering algorithm.

3.1 The Improved Nonlocal Mean Denoising Method

The basic idea of the non-local mean denoising method is to select a fixed-size neighborhood window. It is centered on the pixel which is prepared to be recovered. It searches the image block similar to the central neighborhood window in the entire image range. Assigning the weight with the similarity of the image block, it recovers center point pixel value for all similar blocks basing on average weight. Although the NLM method can effectively remove the noise contained in the texture detail and restore the original image in the field of spatial denoising algorithms. However, when the computational complexity is too high to measure the similarity of image blocks, the weighted average of windows with relatively low similarity may cause partial loss of information such as edges, details, and so on.

The improved nonlocal mean denoising method is based on the super pixel segmentation [5]. It is mainly to optimize the selection of similar windows and put forward a new similar window selection strategy. In general, a region with a flat structure can be considered to have more similarities in the search window, and the majority of the similar blocks obtained by translating a similar window within searching windows have a high similarity. Making these similar windows directly weighted average can acquire better denoising [18]. While structurally rich regions often do not have this feature, they need to explore a more effective similar window selection method. Simultaneously, in different regions of the image with different structural characteristics selecting different pixel sizes of similar windows may also cause similarity measure errors. In regions where the structure is relatively gentle, a similar window with a larger size can be selected to measure the similarity more accurately, and a similar window with a higher similarity can be selected in a similar window with a smaller size in the region where the texture changes. This can optimize the algorithm, retain texture information better, reduce the texture of the fuzzy.

3.2 The Improved Dark Channel Prior Algorithm

The principle of dark channel prior algorithm is that the intensity values of the dark pixels in the underwater image become higher by the filling of the scattered light. By using these dark pixels, the underwater image scattering interference information is estimated to remove these disturbances and get close to true undisturbed underwater images. Although image enhancement by this method can partially remove the noise, the image will have a greater degree of distortion.

The improved dark channel prior algorithm [19, 20] increases the dynamic ranges of the underwater imaging pixels by histogram equalization, transforming the imaging results into YUV space from RGB space for brightness equalization. It can enhance the overall contrast of the image and improve the quality of the image. But also can solve the problem of image discordant resulting by producing singularities when doing equalization in color components of RGB space with original ark channel prior algorithm.

3.3   The Improved Classical Median Filtering Algorithm

The classical filtering algorithm is a non-linear smoothing technique that sets the gray value of each pixel to the median of all pixel points in a neighborhood window. The median filter has a good filtering effect on the impulse noise, especially while filtering the noise, protecting the edges of the signal from blurring. These excellent properties are not available in linear filtering methods. In addition, the median filter algorithm is not only relatively simple, but also easy to make it become true with hardware. Therefore, after the median filter method was proposed, it has applied well in the digital signal processing. However, setting a median value to all pixel gray values will affect the exact judgment of the image.

The improved classical median filtering algorithm is an improved median filtering algorithm based on the combination of noise detection strategy and voting statistics [21]. This algorithm uses the dual-platform histogram equalization algorithm [22] to enhance the distortion image, and then make enhanced images through the adaptive detection of the noise image blocks. Based on the number of pixels according to the gray levels in the filter windows of noise image blocks, it constructs the criteria for voting, which effectively improve the filtering efficiency. Compared with the classical median filtering algorithm, it has the advantages of simple logic and high filtering efficiency [23, 24].

4   Prospects of Image Processing Methods

In recent years, China's deep sea mining level has been improving, and breaking through the depth. However, compared with the developed countries, China's deep-sea mining technology still has a great breakthrough space. As one of the vital part in the development of deep sea mining, deep sea mining video image processing has an excellent improvement prospects.

In improving the underwater image processing methods, we should focus on the three aspects including accuracy, fidelity, efficiency improvement. Obviously, the original classical methods of processing make same treatment to similar regional blocks, images of the different gray-scale value, different dark pixels, but these processes are lack of accuracy. It will lead to some texture blur, incompletely effective denoising and color distortion results. In the modern society with development of science and technology, future underwater image processing methods should be required to become more accurate, effective and maintain its authenticity to a large extent for meeting the needs of deep sea mining, compared to the original rough ranges of processing images achieving a certain degree of denoising, color reproduction and retaining the texture.

References

1. Lu, H., Li, Y., Zhang, L., Serikawa, S.: Enhancing underwater image by dehazing and colorization. Int. Rev Comput. Softw. **7**(7), 3470–3474 (2012)
2. Lu, H., Yamawaki, A., Serikawa, S.: Curvelet approach for deep-sea sonar image denoising, contrast enhancement and fusion. J. Int. Counc. Electr. Eng. **3**(3), 250–256 (2013)
3. Li, Y., Lu, H., Serikawa, S.: Segmentation of offshore oil spill images in leakage acoustic detection system. Res. J. Chem. Environ. **17**(S1), 36–41 (2013)
4. Serikawa, S., Lu, H.: Underwater image dehazing using joint trilateral filter. Comput. Electr. Eng. **40**(1), 41–50 (2014)
5. Lu, H., Li, Y., Zhang, L., Serikawa, S.: Contrast enhancement for images in turbid water. J. Opt. Soc. Am. A **32**(5), 886–893 (2015)
6. Lu, H., Li, Y., Serikawa, S., Li, X., Li, J., Li, K.: 3D underwater scene reconstruction through descattering and color correction. Int. J. Comput. Sci. Eng **12**(4), 352–359 (2016)
7. Lu, H., Li, Y., Hu, X.: Underwater scene reconstruction via image pre-processing. J. Comput. Consum. Control **3**(4), 37–45 (2014)
8. Lu, H., Li, Y., Nakashima, S., Serikawa, S.: Single image dehazing through improved atmospheric light estimation. Multimed. Tools Appl. **75**(24), 17081–17096 (2016)
9. Lu, H., Li, Y., Nakashima, S., Serikawa, S.: Turbidity underwater image restoration using spectral properties and light compensation. IEICE Transactions on Information and Systems, vol. E-99D, no. 1, pp. 219–227 (2016)
10. Lu, H., Hu, X., Li, Y., Li, J., Nakashima, S., Serikawa, S.: Enhancing algorithm based on underwater optical imaging model and guided image filters. J. Yangzhou Univ. **18**(3), 60–63 (2015)
11. Lu, H., Li, Y., Xu, X., Li, J., Liu, Z., Li, X, Yang, J., Serikawa, S.: Underwater image enhancement method using weighted guided trigonometric filtering and artificial light correction. J. Vis. Commun. Image Represent. **36**, 504–516 (2016)
12. Lu, H., Li, B., Zhu, J., Li, Y., Li, Y., He, L., Li, J., Serikawa, S.: Wound intensity correction and segmentation with convolutional neural networks. Concurrency and Computation: Practice and Experience, vol. 29, no. 6 (2017). https://doi.org/10.1002/cpe.3927
13. Lu, H., Li, Y., Nakashima, S., Kim, H., Serikawa, S.: Underwater image super-resolution by descattering and fusion. IEEE Access **5**, 670–679 (2017)
14. Lu, H., Li, Y., Uemura, T., Ge, Z., Xu, X., He, L., Serikawa, S., Kim, H.: FDCNet: filtering deep convolutional network for marine organism classification. Multimed. Tools Appl. 1–10 (2017)
15. Lu, H., Li, Y., Zhang, Y., Chen, M., Serikawa, S, Kim, H.: Underwater optical image processing: a comprehensive review. Mob. Netw. Appl. 1–7 (2017)
16. Lu, H., Zhang, Y., Li, Y., Zhou, Q., Tadoh, R., Uemura, T., Kim, H, Serikawa, S.: Depth map reconstruction for underwater Kinect camera using inpainting and local image mode filtering. IEEE Access 1–9 (2017)

17. Lu, H., Li, Y., Zhang, L., Yamawaki, A., Yang, S., Serikawa, S.: Underwater optical image dehazing using guided trigonometric bilateral filtering. In: Proceedings of 2013 IEEE International Symposium on Circuits and Systems, pp. 2147–2150 (2013)
18. Li, Y., Lu, H., Serikawa, S.: A novel deep-sea image enhancement method. In: Proceedings of 2014 International Symposium on Computer, Consumer and Control, pp. 529–532 (2014)
19. Lu, H., Serikawa, S.: Underwater scene enhancement using weighted guided median filter. In: Proceedings of IEEE International Conference on Multimedia and Expo, pp. 1–6 (2014)
20. Li, Y., Lu, H., Li, J., Li, X., Serikawa, S.: Underwater image enhancement using inherent optical prosperities. In: Proceedings of IEEE International Conference on Information and Automation, pp. 419–422 (2015)
21. Lu, H., Li, Y., Serikawa, S., Li, J., Liu, Z., Li, X.: Image restoration method for deep-sea tripod observation systems in the South China Sea. Proc. MTS/IEEE Ocean. **2015**, 1–6 (2015)
22. Lu, H., Li, Y., Li, X., Xu, X., Mu, S., Nakashima, S., Li, Y., Hu, X., Serikawa, S.: A novel high-turbidity underwater image quality assessment method. In: Proceedings of 2016 International Symposium on Computer, Consumer and Control, pp. 34–36 (2016)
23. Xu, H., Li, Y., Li, Y., Lu, H.: Underwater observing system for Taiping floodgate in Taiping river of Yangzhou. In: Proceedings of 2016 International Symposium on Computer, Consumer and Control, pp. 748–750 (2016)
24. Nakagawa, Y., Kihara, K., Tadoh, R., Serikawa, S., Lu, H., Zhang, Y., Li, Y.: Super resolving of the depth map for 3D reconstruction of underwater terrain using Kinect. In: Proceedings of the 22nd IEEE International Conference on Parallel and Distributed Systems, pp. 1237–1240 (2016)

Key Challenges for Internet of Things in Consumer Electronics Industry

Yukihiro Fukumoto and Kazuo Kajimoto

Abstract The key concept of IoT has the potential to be both a benefit and a threat not only to the ICT industry but also to the consumer electronics and device industries. Panasonic has high expectations for the IoT business and has already started providing value to individuals who live in smart homes. From a business viewpoint, we can simply describe "IoT" as ICT technologies that expanded from the computer to the non-computer world. However, there remain various technological issues to realize a genuine IoT world. This lecture introduces IoT business activities in the consumer electronics industry, and is followed by a discussion of the technological challenges for IoT both in the software and hardware layers.

Keywords IoT · Platform · Data communication · Web of things
EMC (Electromagnetic compatibility)

1 Introduction

Japanese manufacturers have created numerous "T" (terminals, devices) for IoT. The aim is to trigger social innovation by connecting these "Ts" to the "I." Individual IoT devices have in fact already reached a practical level that would enable widespread implementation. Figure 1 shows Panasonic's IoT Future Home (Smart Home) [1]. In Fig. 1a, simply by placing a smartphone over the city's advertisements or lights, you will be directed to stores and provided with product information. Assisting with cooking and other skills with an action function using natural language processing is shown in b. AR and biological sensors that support health and beauty are shown in c. Note that there are no conventional PCs in the picture. What you see are home appliances that look the same as before. However,

Y. Fukumoto (✉) · K. Kajimoto (✉)
Groupwide CTO Office, Panasonic Corporation, Osaka, Japan
e-mail: fukumoto.yukihiro@jp.panasonic.com

K. Kajimoto
e-mail: kajimoto.kazuo@jp.panasonic.com

© Springer International Publishing AG 2018
H. Lu and X. Xu (eds.), *Artificial Intelligence and Robotics*,
Studies in Computational Intelligence 752,
https://doi.org/10.1007/978-3-319-69877-9_25

(a) Wonder Town   (b) Kitchen&Living with dialogue   (c) Relaxation space
                                                                       for health and Beauty

Fig. 1 IoT future home (smart home)

greater value to the customer can be obtained by the system controlled via the Cloud environment through the Internet. IoT technologies are expanding to other applications such as collaboration with automobiles and with the social infrastructure (Smart Towns, Smart Grids).

However, there is concern that the keyword "IoT" is too futuristic to become a buzzword. Whether IoT is simply a term referring to the technical domain or full-scale social innovation depends on whether certain small but important implementation issues can be solved or not at the consumer level. Although implementation tasks are diverse, typical examples are "securing the talent of data analysts," "communication safety and security," "power supply (batteries and WPT)," "reference platforms," "server maintenance management" and so on. These will probably not be tackled by all companies, but they are closely related to the business model and should be solved during the construction of the business ecosystem. In this paper, we describe a part of this subject and discuss its associated challenges.

2   IoT Technology Architecture and Business Development

Information technology has been established using computer and communications technology. IoT extends computer technology to various kinds of electronic elements and it also expands communication into various communication environments and methods. In the IT world, engineers have, over the last 20 years, resolved most computer implementation challenges (Windows, Linux, Intel, and Arm) and communication challenges (TCP/IP, Ethernet, Wi-Fi). In the IoT field, we will need to tackle a mass of implementation problems that are likely to proliferate rapidly over the next few years. In this paper, we describe some examples from the software and hardware aspects of implementation.

Figure 2 illustrates the architecture problems facing IoT software and hardware implementation. The software layer and the cloud layer are parts that are directly connected to the business model of the application. At the platform layer, the struggle for supremacy has triggered intense competition between entities that aim to be the equivalent of Microsoft/Intel in the PC world and Google/Apple/Qualcomm

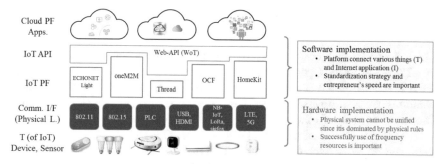

Fig. 2 IoT architecture and its business value

in the smartphone world. On the other hand, the home electronics industry side is the most immediate keychain for developing and standardizing common APIs on the premise of using multiplatforming to avoid the same situation as that which developed in PCs and smartphones, which are tied to specific platforms.

On the other hand, in the hardware layer (phi layer), the characteristic differences caused by physical phenomena overpower any rules set by humans. For example, in wireless systems, low-rate and low-power technologies or high-power and high-rate versions are used depending on the application. The key, therefore, is the ability to determine the method that best takes account of the cost and implementation issues. The increased volume of Internet traffic volume caused by IoT risks conflicts due to the need to use a limited range of frequencies but without causing interference. This is the so-called EMC problem. To avoid this, new technologies such as optical fiber and HD-PLC are being added to designers' options. The ability of the designer to determine the optimal mode of implementation will become increasingly important.

3 Key Challenges in the Software Layer

The key challenges in the software layer are the standardization and construction of reference architecture. In the PC and Smartphone field, the environment for applications (services) is the same regardless of hardware manufacturer, such as Windows for PC, and iOS or Android for smartphones. For this reason, the user can enjoy services provided by various applications, such as document processing, electronic commerce and SNS. However, in IoT, hardware manufacturers individually select their IoT platform, so it is extremely difficult to realize services that combine hardware from different manufacturers [2].

3.1 Silos to Eco Systems

There exist various platforms adopted by hardware manufacturers in IoT, such as one M2M [3], OCF (Open Connectivity Foundation) [4], Echonet Lite [5], Bluetooth Low Energy [6], and so on. The methods and covering hierarchy of these platforms are also diverse. Worse, these platforms are not mutually compatible. Each is fighting to win the race of de facto standardization.

The standardization of the WoT (Web of Things) is currently progressing at the W3C (World Wide Web Consortium). This organization provides a common API of the REST API type that guarantees interoperability among different platforms while taking advantage of the API provided by existing platforms by overlaying a Web technology wrapper on each existing platform group (Fig. 3).

3.2 WoT Building Blocks

WoT assumes that virtual shadow devices of actual equipment exist on the Internet, and each shadow device functions as a micro service for clients. Each shadow device consists of the building blocks shown in Fig. 4 [7].

(1) 'Thing' Description
    Declares the category of the device, attributes, provides an API name, parameter type and so on. The external client refers to this description to call up the WoT Interface.

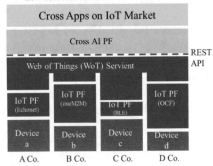

Current Silo Architecture

- No one stop universal PF
- No eco system
- No third party's service
- But services has been launched

WoT Architecture

- One stop universal PF by Web on IoT
- Ecosystem with third party servicers
- **Coexist with other SDO's APIs**
- **Maintain current services as well as new services on WoT**

Fig. 3 Silo architecture versus WoT architecture

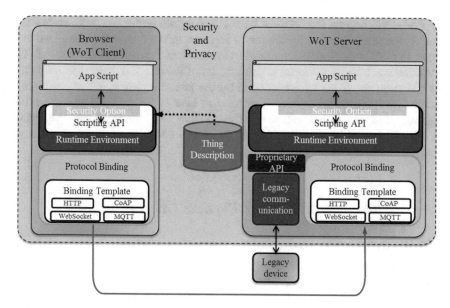

Fig. 4 Concept of WoT building blocks

(2) Scripting API

Implements application logic in a modular and portable way. It can access remote objects through the WoT Interface on the client side. On the server side, it provides register callback entry and so on for creating server logic. For this, the runtime environment provides a Scripting API (Client, Server and Discovery).

(3) Protocol Binding

Converts interactions with devices using information in accordance with lower layer protocols.

(4) Legacy Communication

For legacy devices, adapters available for those devices convert the protocol.

(5) Security and Privacy

The system achieves secure and safe communication controlled by the latest security and privacy technology.

How to design the security and privacy is described in the designing policy document.

4 Key Challenges in the Hardware Layer

While the key challenge in the software layer is standardization and construction of the reference architecture, the main task in the hardware layer is to secure safety and stability.

The world beyond the network layer is a completely digital world, so humans can be designated the decision-makers. Therefore, unless rules are standardized by a de jure or de facto standard, communication between different devices will fail. On the other hand, the physical layer and the part of the data link layer directly linked to it is determined not by humans but by the physical phenomena of the natural world. The physical phenomenon referred to here is the transmission of electric signals, radio waves, light and so on. The designer will therefore address problems such as SI (signal integrity), PI (power integrity), EMC (electromagnetic compatibility), and heat dissipation.

4.1 Various Interfaces in the Physical Layer with Cost Optimization

The key difference between the IoT and the Internet of the PC world is that many parameters, such as cost, data rate, distance, power, and dependability have a variety of values that depend on the object. PCs and smartphones are hardware equipment that cost more than $100 and must have a sufficient power supply. It is inappropriate in terms of power, communication distance, cost, etc., to use them for domestic gas or electricity meters, or as sensors for inspections of dams and bridges. The EMC described in detail in the next section also relates to appropriate use of the radio wave environment. It can be said that EMC is an important technical subject for radio wave usage that is likely to become saturated in terms of frequency and space by the IoT.

Figure 5 shows the home wireless environment for an IoT home, and Table 1 shows the typical physical layer communication method used in IoT. Data rate, transmission distance, transmission cost, power consumption, etc., are in a complicated trade-off relationship, making it necessary to select and combine them optimally to match the application in question.

4.2 EMC: The 'Noise of Things'

The key challenge for the IoT's overall environment is EMC: the interference problem in a limited frequency band. EMC is a rule for allowing equipment that produces EM noise and equipment that receives EM noise to coexist in the same frequency domain and the same space. Most devices and sensors connected to the Internet are noise sources. The number of devices connected to the Internet by the IoT will increase from the current 50 to 10 billion items by a factor of several to dozens by 2020. In short, the IoT will become the NoT: Noise-source of Things, as shown in Fig. 5, creating an environment full of noise sources.

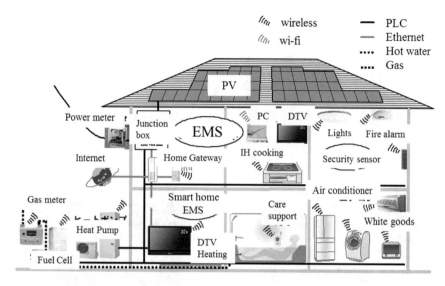

Fig. 5 Data communication in an IoT home

Table 1 Representative example of IoT communication method

Type	Transfer rate	Wireless freq. bands	Feature for IoT use	Example of standard
Ethernet	10 Mbps–100 Gbps	(wired)	Basic method of internet communication	
802.11	2–54 Mbps	2.4, 5 GHz, etc.	Basic wireless method (No license required)	Wi-Fi
802.15	20 kbps–24 Mbps	900 MHz, 2.4 GHz	Short range, low power consumption	ZigBee, Wi-SUN Bluetooth
PLC	210 Mbps	(wired)	Last 1-mile communication using power lines	HD-PLC, UPA, HomePlug AV
LPWA	Sub-GHz	100–250 kbps	Wide range, low-rate low power consumption	LoRa, SIGFOX, Wi-SUN
Cellular	20 kbps–325 Mbps	800 MHz, 1.5 GHz	Service provided by licensed carrier	LTE NB-IOT (LPWA)

Spatially, noise sources in the home are digital equipment, switching power supplies and the like. The number of noise sources in the home has increased 100-fold or more in the past twenty years. It is the same in terms of frequencies. Figure 6 shows the radio frequencies used inside the home. The 2.4 GHz (unlicensed) band is almost saturated, and AV transmissions, which need real-time processing in this band, are becoming extremely difficult [8].

The EMC standard was established by the IEC and CISPR in an era when houses had a much less crowded electromagnetic environment. Therefore, as shown in Fig. 7, interference now occurs in the actual use environment.

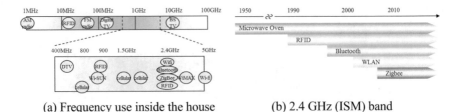

| (a) Frequency use inside the house | (b) 2.4 GHz (ISM) band |

Fig. 6 Frequency domains for the IoT environment

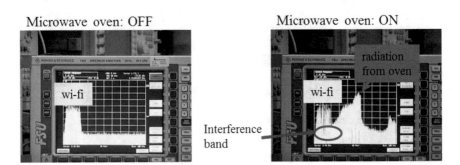

Fig. 7 EM interference in the 2.4 GHz band

In the hardware implementation of IoT, therefore, it will be very important to design a hardware layer that takes the EMC environment into account. Implementation focusing only on individual sensors and individual services is not useful in an actual use environment. Unlicensed radio bands are heading toward saturation, and the selection and utilization of communications environments using wired systems such as Ethernet, PLC, etc., will be important not only in the home environment, but also in offices and factories.

5 Conclusion

IoT entrepreneurs are working to extend the technology very fast, based on the IT infrastructure that took a period of 20 years or more to be established. To be able to control this rush, it is necessary to solve specific implementation issues concerning maintenance of the platform in the software layers and the utilization of frequency resources in the hardware layers.

References

1. Examples of Technologies for IoT/Robotics, http://www.panasonic.com/global/corporate/technology-design/10years-vision/iot-example.html
2. Getting Started with WoT Project @ Osaka F2F, https://www.w3.org/2017/05/wot-f2f/slides/WoT_Getting_started_in_OsakaF2F.pdf
3. oneM2M, http://www.onem2m.org/
4. OCF, https://openconnectivity.org/
5. Echonet Lite, https://echonet.jp/english/
6. Bluetooth Low Energy, https://www.bluetooth.com/
7. Web of Things Architecture, https://w3c.github.io/wot-architecture/
8. Fukumoto, Y., etc.: EMC from both the System and the Component Viewpoints, EMC, No. 300, pp. 13–70 (2013), (in Japanese)

More Discriminative CNN with Inter Loss for Classification

Jianchao Fei, Ting Rui, Xiaona Song, You Zhou and Sai Zhang

Abstract Recently years, convolutional neural networks (CNN) has been a hot spot in various areas such as object detection, classification. As deep study in CNN, its performance is almost human-competitive. We find that the test accuracy largely depends on the relationship of samples in feature space. Softmax loss is widely used in many deep learning algorithms. However, it cannot directly reflect this kind of relationship. In this paper, we design a new loss function, named inter loss. This inter loss function can maximizes the distance between different classes, analogous to maximizing margin in SVM. By integrating inter loss and softmax loss, larger inter-class distance and smaller intra-class distance can be obtained. In this way, we can significantly improve the accuracy in classification. Impressive results is obtained in SVHN and CIFAR-10 datasets. However, our main goal is to introduce a novel loss function tasks rather than beating the state-of-the-art. In our experiments, other forms of loss functions based on inter and intra class distance is also considered as to demonstrate the effectiveness of inter loss.

Keywords CN · Classification · Softmax loss · Inter loss
Loss function

J. Fei · T. Rui (✉) · X. Song · S. Zhang
College of Filed Engineering, PLA University Science and Technology,
Nanjing, China
e-mail: rtinguu@sohu.com; rtinguu3@126.com

J. Fei
e-mail: 349318312@qq.com

X. Song
e-mail: songxiaona1@163.com

S. Zhang
e-mail: 466908114@qq.com

Y. Zhou
Jiangsu Institute of Commerce, Nanjing, China

© Springer International Publishing AG 2018
H. Lu and X. Xu (eds.), *Artificial Intelligence and Robotics*,
Studies in Computational Intelligence 752,
https://doi.org/10.1007/978-3-319-69877-9_26

1   Introduction

Deep learning becomes a hot spot in various fields. Convolutional neural network (CNN) obtains impressive results with its structure design inspired by human visual mechanism [1], such as object detection [2–5], classification [6–8] and segmentation [9–12], et al. Recent years, many researchers get higher accuracy by revise net's inner structure. Hyper-parameters is a optimization direction, such as Xavier [13], Adam [14] and AdaGrad [15]. Another direction is focused on the training process. Srivastava et al. [16] proposes Dropout prevent overfitting; In [17], intermediate supervision address gradient vanishing; Ioffe et al. [18] propose Batch Normalization to alleviate internal covariate shift. Increasing the width and depth of net is also a direction. AlexNet [6] has 8 layers and wins the first place in ILSVRC 2012. VGG net [19] decreases the ImageNet top-5 error to 7.3% with 16 layers, 55% improvement compared with AlexNet.

In this paper, we propose a novel loss function, inter-loss, mainly focusing on the relationship of different classes in training process. The traditional softmax loss function is conceptually identical to a softmax layer followed by a multinomial logistic loss layer offering numerically stable gradient. However, it does not focus on the relationship between different classes directly. Our main goal is to maximize the inter-class distance. We obtain the different square distances between samples of one class and the other class centers. Then we maximize the sum of them. With the joint supervision of inter loss and softmax loss, smaller intra-inter class distance ratio can be obtained. In this way, we can significantly increase the probabilities of correct classification.

2   Related Work

Loss function is aimed to estimate that how much the prediction deviates from its corresponding true value [20]. In classification, loss function is designed to obtain best decision boundaries so that objects can be classified correctly. For example, hinge loss function in SVM [21] obtains larger margins among different categories. Softmax loss function [22] shows excellent classification performance [6, 19, 21, 22]. There are also many other loss functions such as squared loss [23] and Boosting [24].

From probability view, representing a categorical distribution is the main effect of softmax loss function [25]. However, we want to have some other effects in some cases. Integrating softmax loss function with other forms becomes an efficient way to supervise CNN to realize our thoughts. A temporal sparsity term [26] together with logistic loss function, can make each branching component focus on a subset of fine categories. Positive-sharing loss function [27] is a kind of loss caused by incorrect estimation between zero and nonzero labels. During our research, the center loss integrated with softmax loss, minimizing the intra-class variations,is explored in concurrent work by Yandong Wen [28]. The center loss applies constraint on the intra-class distance while our proposed inter loss pay attention to not only the intra-class distance, but also the inter-class distance. Through experiments, inter loss can obtain lower intra-inter class distance ratio and achieve higher accuracy.

3 Additional Information Required from Authors

3.1 Inter Loss

The inter loss is defined as Eq. 1.

$$J_{inter} = \frac{1}{2} \sum_{i=1}^{m} \sum_{j=1}^{n} \sum_{k=1}^{m} \left\| x_i^j - M_k \right\|_2^2 \;, M_k = \frac{\sum_{k}^{q} x_k^p}{q}, \; i \neq k, \tag{1}$$

where n is training batch size, m is the number of class, q is the number of class k. x_i^j is the ith deep feature, belonging to sample j, M_k is the center of class k.

In fact, we should get the square distance between each sample of one class and every other class center. Then consider the sum as inter loss. Larger inter-class distance can be achieved by maximizing inter loss J_{inter}.

So, the whole loss function is:

$$J_{loss} = J_{softmax} + \alpha \frac{1}{J_{inter}}, \tag{2}$$

in which

$$J_{softmax} = - \sum_{i=1}^{m} \sum_{j=1}^{n} \log(P_i^j), \; P_i^j = \frac{\exp(x_i^j)}{\sum_{l}^{m} \exp(x_i^j)}. \tag{3}$$

where $J_{softmax}$ is the softmax loss, ∂ is a scalar to balance softmax loss and inter loss. In order to minimum J_{loss}, we compute the reciprocal value of J_{inter}. As to back propagation, the gradient of loss function and corresponding error term is needed. Here the details of the derivative of J_{inter} is provided:

$$\delta_{inter} = \frac{\partial J_{inter}}{\partial z_i^j} = \frac{1}{2} \sum_{i=1}^{m} \sum_{j=1}^{n} \sum_{k=1}^{m} \frac{\partial}{\partial z_i^j} \left\| x_i^j - M_k \right\|^2 = \sum_{i=1}^{m} \sum_{j=1}^{n} \sum_{k=1}^{m} (x_i^j - M_k) \bullet [(x_i^j)' - (M_k)']$$

$$= \sum_{i=1}^{m} \sum_{j=1}^{n} \sum_{k=1}^{m} (x_i^j - M_k) \bullet (F'(z_i^j)), \quad j \neq k \tag{4}$$

where $F(\bullet)$ is the activation function, z_i^j is the input to the i-th unit in the j-th layer. The derivative process refers to Ufldl tutorial. When we get the error term of last layer, we can updated values for weights and biases layer by layer. In the end, we can complete the whole back propagation.

3.2   Discussion

A simplified CNN model is designed in this section on SVHN, which is similar to [28]. The main difference with the model in Sect. 4.1 is one less convolution layer and 2 hidden layers which is designed for visualization. Figure 1 shows the architecture.

Figure 2 is the distributions obtaining via softmax loss function. Obviously, most samples of same class is clustered together. So we can make the decision boundaries unequivocally. However, here are two concerns:

1. From Fig. 2 we can see that some samples (bad samples) has relatively large intra-class distance, misleading the classification.
2. Obviously, the inter-class distance is not large enough.

As to problem 1, center loss [28] is a way to address it. By center loss, most of the bad examples are around their centers. However, there are always a few of examples that are relatively far from their centers. To solve this issue, we introduce inter loss. Inter loss can apply restrain on inter-intra class distance. In this way, we can obtain smaller intra-inter class distance ratio. In Fig. 3, larger inter class distance is achieved via inter loss.

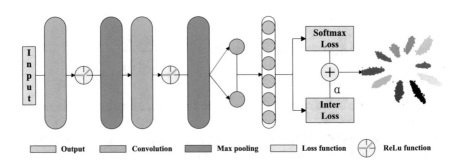

Fig. 1 Architecture of simplified CNN

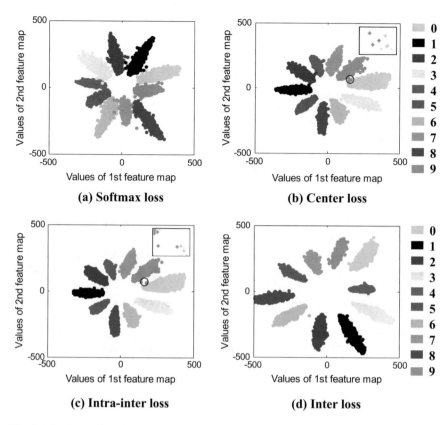

Fig. 2 Distributions of different loss function in 2D

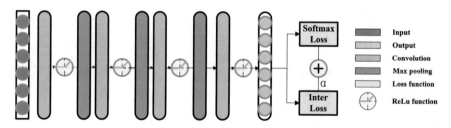

Fig. 3 An illustration of the architecture of our model

4 Experiment

4.1 Model Design

A simple CNN model in LightNet [29] is introduced in our experiments instead of a sophistical one. The overall CNN architecture is shown in Fig. 3.

4.2 Tests on SVHN and CIFAR-10

We use CNN with softmax loss function (model A), center loss function (model B), intra-inter loss function (model C) and inter loss function (model D, proposed) on SVHN and CIFAR-10 dataset.

As to inter loss, hyper parameter ∂ is critical to the performance. We have tested a series of numbers for ∂. The results is shown in Fig. 4. We select ∂ = 1e-2 for SVHN and ∂ = 1e-4 for CIFAR-10.

Next, we have tested the four models on these two datasets. Table 1 is the results.

As shown in Table 1, our proposed model is better than the other models. Its accuracy on and CIFAR is 92.70% and 77.75%, while model with center loss

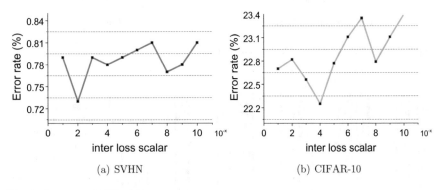

(a) SVHN   (b) CIFAR-10

Fig. 4 Error rate of different

Table 1 Accuracy of different loss on SVHN and CIFAR-10. Data in bracket is the mean accuracy of 10 records

Loss function	Softmax loss	Center loss	Inter loss	Intra-inter loss
SVHN	91.21 (91.25)	92.1 (92.1)	92.70 (92.71)	92.01 (92.05)
CIFAR-10	76.7 (76.6)	77.15 (77.10)	77.75 (77.6)	77.34 (77.30)

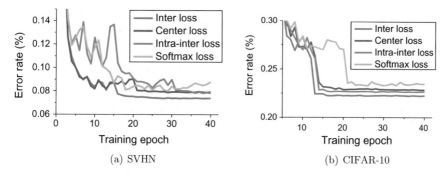

Fig. 5 Error rate of different loss function

achieves 92.1% and 77.15% separately. Intra-inter loss's performance is almost identical to center loss and softmax loss's is worst of all. The curves of accuracy are shown in Fig. 5.

5 Conclusions

Nowadays, the structure improvement and parameter optimizing of CNN models is the main optimization trend. We propose a novel loss function, called inter loss. The inter loss mainly focuses on the relationship between samples. Together by inter loss and softmax loss, distance between samples belonging to different classes is enlarged while distance to their belonging class centers becomes smaller. As a result, inter loss function is versatile to classification. In the end, a series of experiments are conducted to demonstrate its effectiveness.

Acknowledgements This work is supported in part by National Natural Science Foundation of China: NO. 61472444, 61472392.

References

1. Serre, T., Wolf, L., Poggio, T.: Object recognition with features inspired by visual cortex, In: 2005 IEEE Computer Society Conference on Computer Vision and Pattern Recognition (CVPR' 05), vol. 2, IEEE, pp. 994–1000 (2005)
2. Sermanet, P., Eigen, D., Zhang, X., Mathieu, M., Fergus, R., LeCun, Y.: Over-feat: Integrated recognition, localization and detection using convolutional networks. arXiv:1312.6229
3. Girshick, R., Donahue, J., Darrell, T., Malik, J.: Rich feature hierarchies for accurate object detection and semantic segmentation, In: Proceedings of the IEEE conference on computer vision and pattern recognition, pp. 580–587 (2014)
4. He, K., Zhang, X., Ren, S., Sun, J.: Spatial pyramid pooling in deep convolutional networks for visual recognition, In: European Conference on Computer Vision, Springer, pp. 346–361 (2014)

5. Ren, S., He, K., Girshick, R., Sun, J.: Faster r-cnn: towards real-time object detection with region proposal networks, In: Advances in Neural Information Processing Systems, pp. 91–99 (2015)
6. Krizhevsky, A., Sutskever, I., Hinton, G.E.: Imagenet classification with deep convolutional neural networks, In: Advances in Neural Information Processing Systems, pp. 1097–1105 (2012)
7. Lu, H., Li, Y., Uemura, T., et al.: FDCNet: filtering deep convolutional network for marine organism classification. Multimedia tools and applications, 1–14 (2017)
8. He, K., Zhang, X., Ren, S., Sun, J.: Deep Residual Learning for Image Recognition. arXiv:1512.03385
9. Farabet, C., Couprie, C., Najman, L., LeCun, Y.: Learning hierarchical features for scene labeling. IEEE Trans. Pattern Anal. Mach. Intell. 35(8), 1915–1929 (2013)
10. Gupta, S., Girshick, R., Arbeláez, P., Malik, J.: Learning rich features from rgb-d images for object detection and segmentation, In: European Conference on Computer Vision, Springer, pp. 345–360 (2014)
11. Lu, H., Li, B., Zhu, J., et al.: Wound intensity correction and segmentation with convolutional neural networks. Concurr. Computat. Pract. Exper. 29(6) (2017)
12. Long, J., Shelhamer, E., Darrell, T.: Fully convolutional networks for semantic segmentation, In: Proceedings of the IEEE Conference on Computer Vision and Pattern Recognition, pp. 3431–3440 (2015)
13. Glorot, X., Bengio, Y.: Understanding the Difficulty of Training Deep Feed Forward Neural Networks, In: Aistats, vol. 9, pp. 249–256 (2010)
14. Kingma, D., Ba, J.: Adam: a method for stochastic optimization. arXiv:1412.6980
15. Duchi, J., Hazan, E., Singer, Y.: Adaptive subgradient methods for online learning and stochastic optimization, J. Mach. Learn. Res. July 12, 2011 2121–2159
16. Srivastava, N., Hinton, G.E., Krizhevsky, A., Sutskever, I., Salakhutdinov, R.: Dropout: a simple way to prevent neural networks from overfitting. J. Mach. Learn. Res. 15(1), 1929–1958 (2014)
17. Wei, S.-E., Ramakrishna, V., Kanade, T., Sheikh, Y.: Convolutional pose machines. arXiv:1602.00134
18. Ioffe, S., Szegedy, C.: Batch normalization: accelerating deep network training by reducing internal covariate shift. arXiv:1502.03167
19. Simonyan, K., Zisserman, A.: Very Deep Convolutional Networks for Large-Scale Image Recognition. arXiv:1409.1556
20. Christoffersen, P., Jacobs, K.: The importance of the loss function in option valuation. J. Financ. Econ. 72(2), 291–318 (2004)
21. Rosasco, L., De Vito, E., Caponnetto, A., Piana, M., Verri, A.: Are loss functions all the same? Neural. Comput. 16(5), 1063–1076 (2004)
22. Szegedy, C., Liu, W., Jia, Y., Sermanet, P., Reed, S., Anguelov, D., Erhan, V., Vanhoucke, Rabinovich, A.: Going deeper with convolutions, In: Proceedings of the IEEE Conference on Computer Vision and Pattern Recognition, pp. 1–9 (2015)
23. Basu, A., Ebrahimi, N.: Bayesian approach to life testing and reliability estimation using asymmetric loss function. J. Statis. Plann. Infer. 29(1), 21–31 (1991)
24. Freund, Y., Schapire, R.E.: A desicion-theoretic generalization of on-line learning and an application to boosting. In: European Conference on Computational Learning Theory, Springer, pp. 23–37 (1995)
25. Mikolov, T.: Statistical Language Models Based on Neural Networks, Presentation at Google, Mountain View, 2nd April
26. Yan, Z., Jagadeesh, V., Decoste, D., Di, W., Piramuthu, R.: Hd-cnn: hierarchical deep convolutional neural network for image classification, In: International Conference on Computer Vision (ICCV), vol. 2, (2015)
27. Shen, X., Wang, X., Wang, Y., Bai, X., Zhang, Z.: Deepcontour: a deep convolutional feature learned by positive-sharing loss for contour detection, In: Proceedings of the IEEE Conference on Computer Vision and Pattern Recognition, pp. 3982–3991 (2015)

28. Wen, Y., Zhang, K., Li, Z., Qiao, Y.: A discriminative feature learning approach for deep face recognition. In: European Conference on Computer Vision, Springer, pp. 499–515 (2016)
29. Ye, C., Zhao, C., Yang, Y., Fermuller, C., Aloimonos, Y.: Lightnet: A Versatile, standalone Matlab-Based Environment for Deep Learning. arXiv:1605.02766

ConvNets Pruning by Feature Maps Selection

Junhua Zou, Ting Rui, You Zhou, Chengsong Yang and Sai Zhang

Abstract Convolutional neural network (CNN) is one of the research focuses in machine learning in the last few years. But as the continuous development of CNN in vision and speech, the number of parameters is also increasing, too. CNN, which has millions of parameters, makes the memory of the model very large, and this impedes its widespread especially in mobile device. Based on the observation above, we not only design a CNN pruning method, where we prune unimportant feature maps, but also propose a separability values based number confirmation method which can relatively determine the appropriate pruning number. Experimental results show that, in the cifar-10 dataset, feature maps in each convolutional layer can be pruned by at least 15.6%, up to 59.7%, and the pruning process will not cause any performance loss. We also proved that the confirmation method is effective by a large number of repeated experiments which gradually prune feature maps of each convolutional layer.

Keywords Convolutional neural network · Feature maps pruning
CNN simplification · Separability values

J. Zou (✉) · T. Rui (✉) · C. Yang · S. Zhang
PLA Army Engineering University, Nanjing, China
e-mail: zoujhzz@gmail.com

T. Rui
e-mail: rtinguu@sohu.com

T. Rui
State Key Lab for Novel Software Technology, Nanjing University, Nanjing, China

Y. Zhou
Jiangsu Institute of Commerce, Nanjing, China
e-mail: 354442511@qq.com

H. Lu and X. Xu (eds.), *Artificial Intelligence and Robotics*,
Studies in Computational Intelligence 752,
https://doi.org/10.1007/978-3-319-69877-9_27

1 Introduction

Convolutional neural network [1], an important form of deep learning, has become a hotspot in the field of artificial intelligence [2]. CNN is widely used in image classification [3–7], object detection [8–10], saliency analysis [11, 12], target tracking [13] and many other tasks [14–17], and has made a big leap. In training process of CNN, weight sharing, local connection and subsampling operation improve the neural network adaptability and robustness, and greatly reduce the operation time and memory requirement. At the same time, CNN is highly invariant to translation, rotation, tilt, and scaling.

With the constant development of CNN, its performance has been greatly improved, and the depth of CNN (i.e., the layer number of convolutional neural network) is also increasing. As the depth of the network increases, the runtime and memory required to run are also increasing. Previous match results of ImageNet Large Scale Visual Recognition Competition (ILSVRC) show that the depth of the CNN has a positive effect on the performance, but the relationship between the width of the network (i.e., the number of feature maps of each layer) and the performance of CNN has not been discussed. In civilian areas, the storage capacity and computing power of mobile phone are very limited. For example, in the Apple App Store, the application size is usually no more than 200 M, and the number of parameters of commonly used convolutional neural network can be achieved at least 138 million or more on ImageNet dataset. There is also a limitation that we cannot download APP above 100 M in the Apple App Store without connecting Wi-Fi. More than these two issues, CNN is more expensive to calculate. For unmanned systems, the storage capacity and operational performance are limited due to internal space constraints. In order to ensure the performance of the real-time, reducing network complexity and accelerating the running speed become an essential task. The simplification of CNN is a fundamental issue and an important research direction for deep learning, so the solution of the problem is of great significance to the application of machine learning.

In recent years, a great deal of research has been done on the simplification of CNN. By summarizing the literature, the simplification methods are summarized as follows:

1. Replacing complex networks with shallow networks. In simple problems (such as two classification problems), we design a shallow network, which can still keep the network performance, to approximate the complex network, but in the development of deep learning, with the increase of task complexity, shallow network must be eliminated;
2. Binarizing the complex network value. The input, weights and response parameters of complex network can be binarized, and this process not only can reduce the size of the network, but also accelerate the operation speed. But its disadvantages are also very obvious. The network performance will distinctly decrease;

3. Compressing trained complex networks. Quantify the parameters of the network and use other ways to complete the compression of complex networks. However, the extent of compression is still lacking in quantitative analysis and description.

Based on the observation above, we compress the complex network by feature maps pruning which simplifies CNN. Our goal is not only to reduce the weight in the neural network by pruning, but also to find a way to rapidly determine the pruning number of feature maps. It is possible to avoid reducing feature maps in each convolution layer and save a lot of training time by using a valid value determination method. In this paper, we need to prune the feature map of each convolutional layer, design the feature maps filter formula, calculate the separability value of the feature map, define the concept of the critical point, and quickly refer to the relatively small number of reservations of the feature maps. An overview of our model is shown in Fig. 1.

This model can be divided into the following three steps:

1. In CNN training stage, we train a VGG-like network with many 3×3 filters and set padding to 1, and all pooling layers are max pooling. We use both batch normalization and dropout in the network.
2. In feature maps pruning stage, we use the selection formula to calculate separability values of feature maps, and prune feature maps which have small separability values. In the matter of feature map reserve number, we define the concept of critical point which can rapidly refer to relatively small reserve number of feature maps.
3. In CNN retraining stage, we prune feature maps layer by layer from the bottom to the top layer, and retrain CNN until network performance is restored. We repeat these two operations until all convolutional layers are pruned.

The main contribution of our pruning model includes:

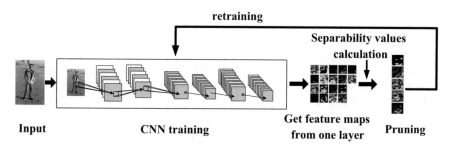

Fig. 1 Three stages of CNN Simplification: training, pruning, retraining. Train a VGG-like network which has achieved the best performance. Prune feature maps which are redundant to network and finish this process layer by layer from the bottom to the top layers. Retrain network after pruning one layer

1. We design a new feature map selection formula. In this paper, we prune CNN by feature maps selection, and we design a new feature map selecting formula to calculate separability values of feature maps. We keep feature maps which have high separability values.
2. We define the concept of critical point. The critical point is an indicator of pruning number. We sort separability values from large to small and build curves. We take the break point of the curve gradient as the critical points.

2 Related Work

Some research on reducing the computing cost and storage by model compression [18–20] is proposed in recent years. Deep compression [18] prunes the network, quantizes the weights and applies Huffman coding to reduce the storage requirement of CNN. Binarized Neural Network [19] trains neural network with weights and activations constrained to +1 or −1 to make shorter runtime. Neural network connections sparsifying [20] proposes a new neural correlation-based weight selection criterion and verifies its effectiveness in selecting informative connections from previously learned models in each iteration, this criterion can stabilize the performance of the baseline model with fewer original parameters.

CNN pruning is preliminarily carried out in these years. Li et al. [21] calculates the absolute sum of kernel weight and prunes filters with the smallest sum values. But this method cannot confirm the best or appropriate pruning number of filters, and the neural network still have many excess filters.

In previous related research we have done before [22, 23], we have proved that CNN can achieve different levels of simplification on different databases or different tasks. But these two feature maps pruning methods are only proved in shallow CNN such as Lenet-5. We need to conduct experiment to prove the theoretical validity of our model in deep neural network such as VGG-16.

3 Feature Maps Pruning Method

In this paper, the core work of CNN simplification is pruning, so we quantify the feature maps of each convolution layer, and thus cut the network intuitively. We design the selection formula as follows:

$$J = tr\left(S_b S_w^{-1}\right) \tag{1}$$

where S_b represents the between-class scatter matrix, S_w represents the within-class scatter matrix, and tr represents the trace of a matrix.

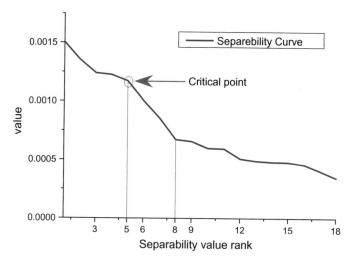

Fig. 2 An example of finding the critical point in separability curve

In feature maps selection formula, the operations of between-class scatter matrix S_b and within-class scatter matrix S_w are the same as linear discriminant analysis (LDA). The difference is that we optimize LDA and use trace to calculate separability values of feature maps, because trace of a matrix is equal to the sum of the eigenvalues of it. We can retain data information as much as possible with the selection formula we designed.

With formula (1), we can quantitatively obtain the separability values of feature maps, but it is not enough for CNN pruning. In deep neural network, such as VGG, there are many convolutional kernels in each convolutional layer. It is not enough that only to find which feature maps to prune. We need to chase down the least feature map reserve number on the premise of stabilizing CNN. Based on the separability of feature map, the separability curve is established, and the breakpoint of the curve gradient is defined as the critical point. An example is given in Fig. 2.

4 Experiments

Experiment is carried out on Cifar-10 dataset with VGG-16. Follow the pruning steps we proposed in Fig. 1, we firstly deal with the first convolutional layer (Conv_1). There are 64 convolutional kernels in Conv_1, so we can get 64 separability values of 64 groups of feature maps. The separability curve of Conv_1 is shown in Fig. 3. We indicate 5 critical points to prevent least reserve number omission.

After trying the 5 critical points marked in Fig. 3, we confirm the appropriate feature map reserve number of Conv_1 is 30 while 64 in VGG-16. Based on the

Fig. 3 Separability curve of Conv_1

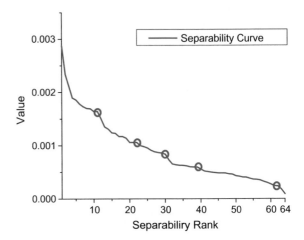

result of Conv_1, the number of kernels of the second convolutional layer (Conv_2) should be more than 30, so the critical points of Conv_2 is at least 30. The following work are the same as Conv_1, marking the critical points of the separability curve of Conv_2, trying the critical points, determining appropriate reserve number and retraining the network. Finally, feature maps number of the Conv_2 is 46. Based on the result of Conv_2, the number of kernels of the third convolutional layer (Conv_3)should be more than 46, so the critical points of Conv_3 is at least 46. Subsequent experiments are carried out according to this rule. After all the

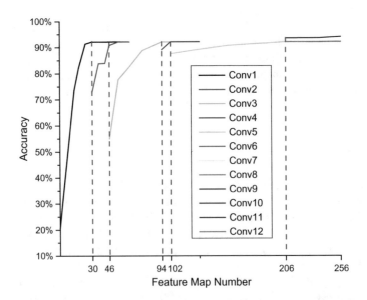

Fig. 4 Accuracy versus feature map reserve number under the screening formula we designed

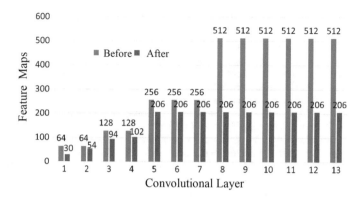

Fig. 5 The comparison of feature map number between before and after simplification

convolution layers are pruned, the number of feature maps of the first layer to the fourth layer is respectively 30, 46, 94, and 102, and the number of feature maps is 206 for the leftover layers. The experimental results are shown in Fig. 4.

Figure 5 shows the comparison of feature map number between before and after pruning. As can be seen, feature maps in low layers can only prune in a low proportion, and the lowest reaches 15.6%. Feature maps in high layers can prune in a low proportion, and the highest reaches 59.7%.

5 Conclusion

In this paper, we present a CNN pruning method by feature maps selection which can reduce the number of feature maps on the premises of stabilizing model performance, and the feature maps we prune can be confirmed by the separability values. We also define the concept of critical point which refer to appropriate feature map number of each layer, and the critical point is the break point of the curve gradient. In this method, we can not only reduce the in-memory footprint of CNN, but also shorten the running time by pruning feature maps.

Acknowledgements This work was supported in part by National Natural Science Foundation of China (GrantNo.61472444, 61472392).

References

1. LeCun, Y., Bengio, Y., Hinton, G.: Deep learning. Nature **521**(7553), 436–444 (2015)
2. Lu, H., Li, Y., Chen, M., et al.: Brain intelligence: go beyond artificial intelligence (2017)
3. Zeiler, M.D., Fergus, R.: Visual. Understand. Convolution. Netw. **8689**:818–833 (2014)

4. Krizhevsky, A., Sutskever, I., Hinton, G.E.: ImageNet classification with deep convolutional neural networks. In: International Conference on Neural Information Processing Systems. Curran Associates Inc., 1097–1105 (2012)
5. Sermanet, P., Eigen, D., Zhang, X., et al.: OverFeat: integrated recognition, localization and detection using convolutional networks. Eprint Arxiv (2013)
6. Chatfield, K., Simonyan, K., Vedaldi, A., et al.: Return of the devil in the details: delving deep into convolutional nets. Comput. Sci. (2014)
7. Li, Y., Lu, H., Li, J., et al.: Underwater image de-scattering and classification by deep neural network. Comput. Electr. Eng. **54**, 68–77 (2016)
8. Xu, W., Xu, W., Yang, M., et al.: 3D Convolutional neural networks for human action recognition. IEEE Trans. Pattern Anal. Mach. Intell. **35**(1), 221–231 (2012)
9. Zou, W.Y., Wang, X., Sun, M., et al.: Generic object detection with dense neural patterns and regionlets. Eprint Arxiv (2014)
10. Liu, J., Ren, T., Wang, Y., et al.: Object proposal on RGB-D images via elastic edge boxes. Neurocomputing (2016)
11. Guo, J, Ren, T., Bei, J., et al.: Salient object detection in RGB-D image based on saliency fusion and propagation. In: International Conference on Internet Multimedia Computing and Service. ACM, **59** (2015)
12. Ren, T., Liu, Y., Ju, R., et al.: How important is location information in saliency detection of natural images. Multimed. Tools Appl. **75**(5), 2543–2564 (2016)
13. Ren, T., Qiu, Z., Liu, Y., et al.: Soft-assigned bag of features for object tracking. Multimed. Syst. **21**(2), 189–205 (2015)
14. Razavian, A.S., Azizpour, H., Sullivan, J., et al.: CNN features off-the-shelf: an astounding baseline for recognition. In: IEEE Conference on Computer Vision and Pattern Recognition Workshops. IEEE Computer Society. 512–519 (2014)
15. Taigman, Y., Yang, M., Marc, et al.: DeepFace: closing the gap to human-level performance in face verification. Comput. Vis. Patt. Recogn. IEEE, 1701–1708 (2014)
16. Lu, H., Li, B., Zhu, J., et al.: Wound intensity correction and segmentation with convolutional neural networks. Concurr. Computat. Pract. Exper. (2016)
17. Lu, H., Li, Y., Nakashima, S., et al.: Single image dehazing through improved atmospheric light estimation. Multimed. Tools Appl. **75**(24), 17081–17096 (2016)
18. Han, S., Mao, H., Dally, W.J.: Deep compression: compressing deep neural networks with pruning, trained quantization and huffman coding. Fiber **56**(4), 3–7 (2015)
19. Courbariaux, M., Bengio, Y.: BinaryNet: Training Deep Neural Networks with Weights and Activations Constrained to +1 or −1 (2016)
20. Sun, Y., Wang, X., Tang, X.: Sparsifying neural network connections for face recognition. In: IEEE Conference on Computer Vision and Pattern Recognition. IEEE Computer Society, 4856–4864 (2016)
21. Li, H., Kadav, A., Durdanovic, I., et al.: Pruning Filters for Efficient ConvNets (2017)
22. Rui, T., Zou, J., Zhou, Y., et al.: Pedestrian detection based on multi-convolutional features by feature maps pruning. Multimed. Tools Appl. 1–11 (2017)
23. Rui, T., Zou, J., Zhou, Y., et al.: Convolutional neural network feature maps selection based on LDA. Multimed. Tools Appl. 1–15 (2017)

The Road Segmentation Method Based on the Deep Auto-Encoder with Supervised Learning

Xiaona Song, Ting Rui, Sai Zhang, Jianchao Fei and Xinqing Wang

Abstract The environment perception of road is a key technique for unmanned vehicle. Determining the driving area through segmentation of road image is one of the important methods. The segmentation precisions of the existing methods are not high and some of them are not real-time. To solve these problems, we design a supervised deep Auto-Encoder model to complete the semantic segmentation of road environment image. Firstly, adding a supervised layer to a classical Auto-Encoder, and using the segmentation image of training samples as the supervised information, the model can learn the features useful for segmentation to complete the semantic segmentation. Secondly, the multi-layer stacking method of supervised Auto-Encoder is designed to build the supervised deep Auto-Encoder, because the deep network has more abundant and diversified features. Finally, we verified the method on CamVid. Compare with CNN and FCN, the road segmentation performances such as precision, speed are improved.

Keywords Image segmentation · Road recognition · Auto-Encoder
Semantic segmentation · Unmanned vehicle

X. Song · T. Rui (✉) · S. Zhang · J. Fei · X. Wang
College of Filed Engineering, PLA University Science and Technology, Nanjing, China
e-mail: rtinguu@sohu.com

X. Song
e-mail: songxiaona1@126.com

S. Zhang
e-mail: 466908114@qq.com

J. Fei
e-mail: feijianchao2010@163.com

X. Wang
e-mail: ruwaye@126.com

X. Song
College of Mechanical Engineering, North China University of Water Resources
and Electric Power, Zhengzhou, China

© Springer International Publishing AG 2018
H. Lu and X. Xu (eds.), *Artificial Intelligence and Robotics*,
Studies in Computational Intelligence 752,
https://doi.org/10.1007/978-3-319-69877-9_28

1   Introduction

The road environment perception of unmanned vehicle has always been a hot research area, and the method based on machine vision is one of the important research methods. The process of the traditional image segmentation is quite complex, and the result is often unsatisfactory. As a classical model of deep learning, Auto-Encoder can learn the important features from samples by unsupervised self-learning and reconstruct the data information through concise expression. The supervised deep Auto-Encoder is designed in this paper and is applied on the semantic segmentation of the road environment successfully.

The contributions of this paper are as follows:

(1) A new semantic segmentation method is proposed in this paper. Adding a supervised layer to a classical Auto-Encoder to learn useful features for image segmentation, we complete the road image segmentation.
(2) Because of the supervised layer of Auto-Encoder, the traditional stacking method is unsuitable. We design the multi-layers stacking method for supervised Auto-Encoder and build a supervised deep Auto-Encoder which has better performance of feature extraction than single-layer supervised Auto-Encoder.
(3) It is proved that the road segmentation performance of the method proposed is effective and simple on the Camvid. And compare with other methods, the segmentation precision of road region is better than others.

2   Related Work

There are many methods for road environment perception, and the methods using the deep learning [1–4], have been a research hotspot. Brust et al. [5] use the CNN to learn features of road environment. An image patch is classified by CNN and the probability which belongs to sky, surroundings and road is given to complete the road segmentation. This method which detects road of the whole image by sliding window is not real-time. Long et al. [6] propose the semantic segmentation method based on FCN which transform fully connected layers into convolution layers, and it enables a classification net to output a heatmap and get the semantic segmentation through upsampling. Then the upsampling of network is improved by Badrinarayanan [7] who build a convolutional encoder-decoder architecture by the name of Segnet. The performance of Segnet is tested on CamVid and other datasets, and it achieves good segmentation performance. Although the methods get good segmentation results, the model of FCN is too complex, and the fine tuning of parameters is a lengthy task, and it needs long time to train the model and also needs high-performance system hardware. All the problems mentioned above affects its application on road detection.

As a classical model in deep learning, Auto-encoder [8, 9], boosts efficiency through automatic feature extraction and shows the ability to learn the essential features from numerous unlabeled samples and reconstruct the initial image through concise and effective expression. Vincent [10] presents the Stacked Denoising Auto-Encoder which can learn feature of data and reconstruct the initial image. Huang [11] designs an adaptive deep supervised Auto-Encoder to reconstruct the damaged face and got a good reconstruction result. The Convolutional Auto-Encoder is used to smooth the image to reduce the difficulty of unsupervised image segmentation and improve the segmentation precision by Wang [12]. Because the classical Auto-Encoder just can reconstruct the initial image, few researches use Auto-Encoder to complete the image segmentation. This paper propose supervised deep Auto-Encoder model and apply it on the semantic segmentation of road environment.

3 The Proposed Method

A supervised layer is added to the traditional Auto-Encoder, and the segmentation image of road environment is used as the supervised information. Because of the supervised layer, Auto-Encoder can learn features useful for image segmentation and complete the semantic segmentation of the road environment. The experiments of researchers shows that deep networks can learn more abstract and diversified feature, so a deep network is built to extract deep features of the road environment to complete the semantic segmentation. Because a supervised layer is added to a classical Auto-Encoder, the multi-layer stacking form is not applicable. The multi-layers stacking form of supervised model is studied and a supervised deep Auto-Encoder model is designed. It is proved that the performance of supervised deep Auto-Encoder is superior to the single-layer one. As shown in Fig. 1, a supervised deep Auto-Encoder is trained using the training samples and their segmentation image. Testing samples are fed to the model to get the semantic segmentation images which are handled to get the driving area by the general image processing methods.

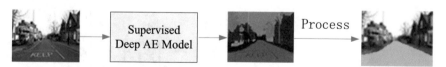

Fig. 1 The proposed method

4   Supervised Auto-Encoder

4.1   Supervised Single-Layer Auto-Encoder

In order to use the AE to complete semantic segmentation, we present adding a supervised learning layer to the classical unsupervised AE model. The objective function of the new model is changed from the mean reconstruction error between the input data X and the reconstructed data Z to the mean reconstruction error between the supervised label X_{label} and the reconstructed sample Z

$$J_{supervised}(W,b) = \frac{1}{m} \sum_{i=1}^{m} (\frac{1}{2} \|Z^i - X_{label}^i\|^2) \tag{1}$$

The existing road segmentation images are used as X_{label}. The supervised AE model tries to learn features useful for segmentation and completes the semantic segmentation for road environment by minimizing the mean reconstruction error between the supervised label X_{label} and the reconstructed sample Z. The architecture of the supervised single-layer AE model is shown as Fig. 2.

In order to learn feature expression, we calculate the residuals of each node from the output layer and update the weights of encoder and decoder by back propagation algorithm. And we use the gradient descending method to update the feature weights under the framework of the supervised learning.

4.2   Supervised Deep Auto-Encoder

The deep structure of the classical AE is achieved through an unsupervised greedy layer-wise training, which is implemented in a stacking form. Each layer of the training process will produce a feature expression. The deeper the number of layers stacked, the more abstract features are obtained. For the supervised deep AE model,

Fig. 2 The supervised single-layer AE model

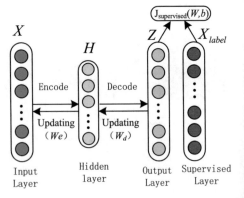

since the supervised label which is used to supervise each layer of the deep AE model is not the input of this layer, the classical AE stacking method and the training strategy can not be used. In this paper, we propose a new deep network structure and training method. The training of the new model is divided into three groups, as shown in Fig. 3. Each group of training is based on a supervised learning single-layer AE model, and each training model is independent, but the entire training process is interrelated. In the training process, we describe the result of the hidden layer of the ith model as the feature code H_i, and the connection weight (W_e^i) from the input layer to the hidden layer is called the encoded weight, the connection weight (W_d^i) from the hidden layer to the output layer is called the decoded weight.

In the first group model, the initial road environment image X is used as the input X_1, and the road segmentation image X_{label} is used as the supervised information, the reconstruction data of the output is described as $\hat{X}_1$. Through supervised learning, the model try to minimize the reconstruction error between the reconstruction data $\hat{X}_1$ and the segmentation image X_{label} through supervised learning. The architecture of the model is shown in Fig. 3a. In the second group model, the reconstruction data $\hat{X}_1$ of the first group is used as input X_2, $\hat{X}_1$ together with road segmentation image X_{label} are used to learn the features and try to decrease the reconstruction error between the data $\hat{X}_2$ and the segmentation image X_{label}. The architecture of the model is shown in Fig. 3b. In the third group model, the feature code H_1 learned from the first group is used as the input X_3, the feature code H_2 learned from the second group is used as the supervision. Through supervised learning, the model get the reconstruction data $\hat{X}_3$, as shown in Fig. 3c.

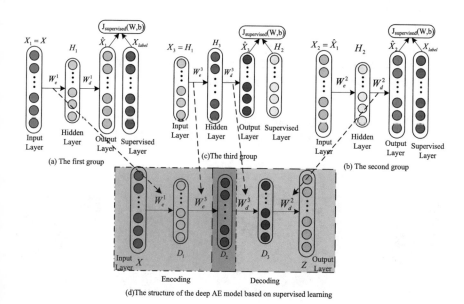

(a) The first group

(b) The second group

(c)The third group

(d)The structure of the deep AE model based on supervised learning

Fig. 3 Training framework of deep AE model based on supervised learning

In the training of each group of supervised single-layer AE models, the sigmoid function is used as an activation function to map the feature representation. According to the training method of supervised single-layer AE in 4.1, we complete the training of three group models. On the basis of the coding and decoding network structure, a stacking AE model is constructed with two layers coding and two layers decoding. The network structure is shown in Fig. 3d. The input of the supervised deep AE model is the initial road image X which is the input of the first group model. Then we extract the encoder weight of the first group W_e^1, the encoder weight of the third group W_e^3, the decoder weight of the third group W_d^3, the decoder weight of the second group W_d^2 as the connecting weights of the subsequent layer. Finally we get the output of the supervised deep AE model Z.

5   Experiments and Analysis

We verified the effectiveness of the proposed method on the Cambridge-driving Labeled Video Database (CamVid). There are 701 pixel-wise semantic segmentation images. We extracted 600 samples and mirrored them horizontally to get 1200 training samples. Then, We extracted 100 samples of the rest images and mirrored them horizontally to get 200 testing samples. In order to verify the effectiveness of the algorithm, we down sampled the image from 540 * 720 to 60 * 80. We trained the three models one by one according to the training process of the supervised deep AE in 4.2 and stacked the three models, and got the supervised deep model shown as Fig. 4.

5.1   Comparison of the Results Between the Supervised Single-Layer AE and Supervised Deep AE

The difference of the image segmentation ability between supervised single-layer AE model and the deep one is studied in this paper. In general, the segmentation

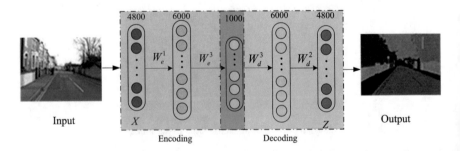

Fig. 4 Supervised deep AE model

image of the deep model is more distinct. This is because the deep and abstract features are extracted in supervised deep AE model, and the boundaries of the each region are clear. The smaller classes such as street lamps, trees, pedestrian, are segmented more clearly. In Fig. 6, each row from top to bottom are testing sample, the results of supervised single-layer AE model, the results of supervised deep AE model and the grey-scale maps of ground truth segmentation. When the iterations of the two models are the same, the segmentation of the deep model is more distinct and less of the noisy point, the quality of segmentation image is better (Fig. 5).

We evaluated the results of the segmentation images on the three metrics: the inter-regional gray-level contrast (GLC), the mean intersection over union (Mean IU) and the intra-regional uniformity (UM), We compute:

$$GLC = \frac{|f_1 - f_2|}{f_1 + f_2} \qquad IU = \frac{S_i \cap G_i}{S_i \cup G_i}$$

$$UM = 1 - \frac{1}{C}\sum_i \{ \sum_{(x,y) \in R_i} [f(x,y) - \frac{1}{A_i} \sum_{(x,y) \in R_i} f(x,y)]^2 \} \qquad (2)$$

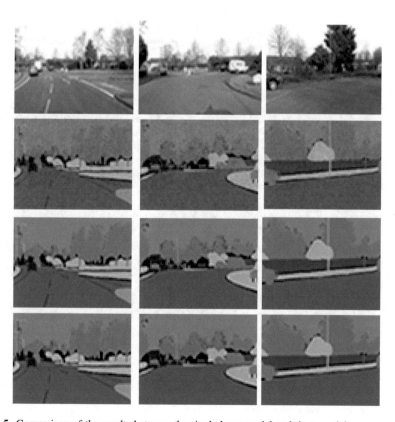

Fig. 5 Comparison of the results between the single-layer model and deep model

Table 1 Comparison of the metrics between supervised single-layer and supervised deep model

Method	GLC	UM	Mean IU (%)
Supervised single-layer AE	0.154	0.634	89.5
Supervised deep AE	0.183	0.857	95.6

The GLC and UM of the deep model is superior to the single-layer model, and the Mean IU of the deep model is up 4.1% from the single-layer model. This demonstrates that the supervised deep AE can extract more deep features of segmentation and excels the single-layer model at the performance of segmentation (Table 1).

5.2 Results and Analysis of the Supervised Deep AE

The 200 testing samples are input to the supervised deep AE model trained by training samples. We get the segmentation results and mark the road region in green in the initial images, as shown in Fig. 6. Each row from top to bottom are testing sample, the segmentation image of the supervised deep AE, the marked road image,

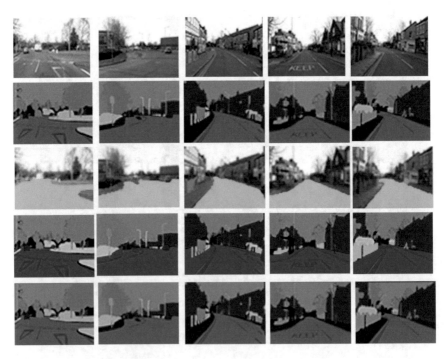

Fig. 6 Results of some testing sample

the ground truth of segmentation, the grey-scale map of the ground truth. We can see that the segmentation results of the supervised deep AE model is almost the same as the grey-scale map of the ground truth from the second row and the fifth row. The model can ignore the details of the image useless for segmentation and retain the boundary of regions in image useful for segmentation. It demonstrates that the model could learn features useful for segmentation through supervised learning and complete the segmentation of the image very well. The algorithm proposed also can detect the vehicles and pedestrian effectively and the boundaries between sidewalk and road are retained completely.

We compare the detection results of the road region with other methods in quantity, as shown in Table 2. The evaluation metrics are calculated as follows:

$$TPR = \frac{TP}{TP+FN} \quad FPR = \frac{FP}{FP+TN} \quad IU = \frac{TP}{TP+FP+FN} \quad (3)$$

where TP is the true positive, TN is the true negative, FP is the false positive, FN is the false negative, and the inference time of every method are also given in Table 2.

Comparing with the Kmeans method, the TPR of the Supervised Deep AE goes up by 14.5% and the FPR goes down 12.5% which demonstrate the super performance of segmentation of the model. The TPR and the inference time are superior to the CNN detection method proposed in [6]. The TPR is 3.8% higher than Segnet. We analyzed the results of Segnet, found that the segmentation precisions of large objective such as road, sky, buildings were lower than others and the segmentation precisions of small objective such as cars, pedestrian, tree are higher than others. It's related to the size of convolutional kernel in the network. For the large region such as road, small convolutional kernel can't learn the holistic features such as the shape of the road, but it can learn the partial features such as color and texture. Because of the illumination, damage and repair, the color and texture features of different region in the same road may be different. But the supervised learning of the method proposed in this paper make the network to learn the contours and boundaries of different region and neglect the image details useless for segmentation, so the method get the better segmentation of road region. Furthermore, the model is simpler, and the training time is shorer and the inference time which is very critical for road detection is far below the Segnet.

Table 2 Results of the road detection between different methods

Method	TPR (%)	FPR (%)	IU (%)	Inference time (s)
Kmeans	78.6	15.8	63.5	0.25
CN24 [5]	95.2	3.5	92.8	1.8
Segnet [7]	93.3	2.8	91.7	0.15
Supervised deep AE	97.1	3.3	95.4	0.03

6   Conclusion

The supervised deep Auto-Encoder model is proposed in this paper and is used to the semantic segmentation of road environment. As is proved in the experiments, compare with Kmeans, CNN and FCN, the algorithm proposed improves the precision and speed of the road segmentation. And supervised deep Auto-Encoder model is simple and effective.

Acknowledgements   This work was supported in part by National Natural Science Foundation of China (GrantNo. 61473444, F011305).

References

1. Mohan, R.: Deep deconvolutional networks for scene parsing. Computer Science (2014)
2. Girshick, R., Donahue, J., Darrell, T., Malik, J.: Rich feature hierarchies for accurate object detection and semantic segmentation. CVPR (2014)
3. Liu, J., Liu, B., Hanqing, L.: Detection guided deconvolutional network for hierarchical feature learning. Pattern Recogn. **48**, 2645–2655 (2015)
4. Oliveira, G.L., Burgard, W., Brox, T: Efficient deep models for monocular road segmentation. In: International Conference on Intelligent Robots and Systems, pp. 4885–4891 (2016)
5. Brust, C.-A., Sickert, S., Simon, M., Rodner, E., Denzler, J.: Convolutional patch networks with spatial prior for road detection and urban scene understanding. VISAPP (2015)
6. Long, J., Shelhamer, E., Darrell, T.: Fully convolutional networks for semantic segmentation. CVPR (2015)
7. Badrinarayanan, V., Kendall, A., Cipolla, R.: SegNet: a deep convolutional encoder-decoder architecture for image segmentation. PAMI (2017)
8. Pathak, D., Kra¨henbu¨hl, P., Donahue, J., Darrell, T., Efros, A.A.: Context encoders: feature learning by inpainting (2016)
9. Masci, J., Meier, U.: Dan Cire¸san, and Ju¨rgen Schmidhuber: stacked convolutional auto-encoders for hierarchical feature extraction. Int. Conf. Artif. Neural Netw. **6791**, 52–59 (2011)
10. Vincent, P., Larochelle, H., Lajoie, I., Bengio, Y., Manzagol, P.-A.: Stacked denoising autoencoders: learning useful representations in a deep network with a local denoising criterion. J. Mach. Learning Res 3371–3408 (2010)
11. Huang, R., Liu, C., Li, G., Zhou, J.: Adaptive deep supervised autoencoder based image reconstruction for face recognition. Math. Probl. Eng. (2016)
12. Wang, C., Yang, B., Liao, Y.: Unsupervised image segmentation using convolutional autoencoder with total variation regularization as preprocessing. ICASSP (2017)

Data Fusion Algorithm for Water Environment Monitoring Based on Recursive Least Squares

Ping Liu, Yuanyuan Wang, Xinchun Yin and Jie Ding

Abstract In recent years, Wireless Sensor Networks (WSNs) has been successfully applied to the water environment monitoring field. But due to the large area of the monitored waters, the great number of sensor nodes and the vast amount of information collected, the redundancy of data is easy to cause network congestion. In these circumstances, data fusion is essential to WSNs-based water environment monitoring system. Data fusion reduces the energy consumption of communications, but at the same time increases the computational energy consumption. For the purpose of saving energy consumption and prolonging network lifetime, it is necessary and significant to study how to reduce the computation complexity of data fusion. This paper establishes a water environment monitoring network model and a data fusion model in the cluster. On the basis of recursive least squares, the forward and backward recursive algorithms are proposed in order to reduce the computation complexity of data fusion, and the advantages of the new algorithms are analyzed in detail.

Keywords Water environment monitoring · Wireless sensor networks
Data fusion · Recursive least squares

1 Introduction

The establishment of a real-time and effective water environment monitoring system is of great significance to the flood and drought control, water pollution control and water ecological restoration in urban areas [1]. Internet of Things (IoT) plus

P. Liu (✉) · J. Ding
School of Information Engineering, Yangzhou University, Yangzhou, China
e-mail: yzuliuping@sina.com

Y. Wang
Yangzhou Polytechnic College, Yangzhou, China

X. Yin
Guangling College, Yangzhou University, Yangzhou, China

© Springer International Publishing AG 2018
H. Lu and X. Xu (eds.), *Artificial Intelligence and Robotics*,
Studies in Computational Intelligence 752,
https://doi.org/10.1007/978-3-319-69877-9_29

water environment monitoring technology brings about a far-reaching and wide range of innovations [2], thus becomes a new development trend in water environment monitoring technology.

With multi-parameter water environment sensors becoming increasingly inexpensive, reliable and low-power, it is entirely possible to deploy these sensors in many rivers and lakes within a limited budget. Owing to the low cost, automation, distributed, high precision and space-time continuity, Wireless Sensor Networks (WSNs) has been successfully applied to a variety of experiments related to water environment monitoring, such as Critical Zone Observatories (CZO) [3], National Ecological Observatory Network (NECO) [4], Long-Term Ecological Research Network (LERN) [5], Heihe Water Allied Telemetry Experimental Research (HiWATER) [6] and so on. The use of WSNs can be an effective way to overcome shortcomings like poor real-time, excessive cost, small monitoring area and so on in the traditional water environment monitoring methods.

Data fusion is very necessary in the WSNs-based water environment monitoring system [7]. The energy consumed by the sensor nodes when transmitting information is much higher than the energy consumed when calculating the data [7, 8]. The data is fused to reduce the amount of data sent and received, which can effectively save the energy of the whole network. Most of the research is also focused on this aspect [9–15]. However, in some applications that require a lot of computing, such as streaming media, video Monitoring, and image-based tracking applications, the data fusion process also needs to consume a certain amount of energy. In these applications, data fusion reduces the energy consumption of communications, but at the same time increases the computational energy consumption. Therefore, reducing the computation complexity of data fusion to save network energy has an excellent value.

2   Network Model of Water Environment Monitoring

In the applications of water environment monitoring, the waters are often vast and widely distributed and the number of sensor nodes is large. The sensor nodes in a WSN are usually clustered, with each cluster having a cluster head. Whether the data fusion technology is used in water environment monitoring network is related to the distance between cluster head and sink node. When the cluster head is far from the sink node, the amount of data consumed by the data fusion is less than that consumed by transmitting redundant data. When the distance is relatively close, the data can be sent directly to the sink node.

Water environment monitoring network often requires real-time and continuous monitoring in a large area of water. The cluster head nodes are usually far from the sink node. Assuming that the routing layer in water environment monitoring network uses multi-path and multi-hop transmission strategy based on Low Energy Adaptive Clustering Hierarchy (LEACH) protocol, the parameters of water environment collecting by the sensor nodes in the waters will be sent to the cluster head

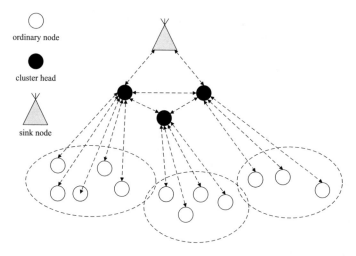

Fig. 1 The transmitting model of water environment monitoring network based on LEACH

nodes. Data are fused and forwarded to the sink node according to the routings between the cluster head nodes. The transmitting model of water environment monitoring network is shown in Fig. 1.

3 Data Fusion Model in the Cluster Based on Least Squares

In the WSNs-based water environment monitoring system, the total number of water environment monitoring sensor nodes is n. There are N sensors in parts of the monitored waters to measure the water environment parameters θ (such as pH, dissolved oxygen, ammonia nitrogen, total nitrogen, total phosphorus, etc.). The N sensors belong to the same cluster. In the stage of stable phase, the collected water environment data are transmitted from ordinary nodes to the cluster head in turns, and then sent to the sink node by the cluster head. Since the N sensors belong to the same cluster, the data collected are highly correlated. If the cluster head is far away from the sink node, sending data to the sink node after data fusion can save a lot of energy and will not bring the parameters θ of the monitored waters a greater loss.

The cluster head establishes a Time Division Multiple Access (TDMA) time slot table which divides one period of time into N time slots and allocates each ordinary node a different time slot. The ordinary node $i(i = 1, 2, \ldots, N)$ transmits the collected data $y_i = h_i\theta + e_i(i = 1, 2, \ldots, N)$ to the cluster head, where e_i follows the N $(0, 1)$ distribution (standard normal distribution) and is independent of each other. And h_i here is the known parameter of the ith measuring device. Owing to sensor accuracy problems, the observations contain a lot of errors in the process of data collecting by water environment monitoring terminals. In fact, what the ordinary node i really collects are data like $h_i \theta + e_i$.

After the N slots, the data received by the cluster head are $y = (y_1, y_2, \ldots, y_N)^T$, which can be characterized by the following linear model:

$$y = H\theta + e, e \sim N(0, \sigma^2 I) \tag{1}$$

In formula (1), H is an $N \times d$ matrix, assuming that it is column full rank, y is the observation vector that has functional relation to θ, θ is the water environment parameter vector, and e is an N-dimensional vector of measurement noise.

e makes the following assumptions:

1. The internal noise of sensor and the measurement noise caused by environmental interference are independent of each other.
2. The measurement noise of each sensor is white Gaussian noise.
3. The distribution rule of measurement noise follows the normal distribution.

Among the theories and methods of linear model parameter estimation, least squares holds the central and basic position [16]. The advantage of least squares is that the optimization rules are very simple and no prior statistical information about the estimated parameters and the errors is needed, so it is very convenient for practical applications. We use least squares estimation method in this paper to fuse the inner-cluster data. On the basis of y and $(H^T H)^{-1} H^T$, the least squares estimation of θ can be obtained by simple operation:

$$\widehat{\theta} = (H^T H)^{-1} H^T y. \tag{2}$$

The cluster head then sends θ to the sink node. According to the Gauss-Markov theorem [16], $\widehat{\theta}$ is the Best Linear Unbiased Estimate (BLUE) of θ. Therefore, using $\widehat{\theta}$ instead of the collected N sets of data will not bring the parameters θ a greater loss in this round.

4 Data Fusion Algorithm Based on Recursive Least Squares

4.1 The Statement of Problem

The data fusion algorithm is responsible for responding to the user's query request. After sending the query request by the user, the sink node requests the inner-cluster fusion result from the cluster head nodes in a certain time interval and the cluster head nodes upload the approximate result to the sink node. At last, data fusion between clusters is performed by the sink node in order to obtain the final data fusion result.

In the water environment monitoring WSNs, for its complex working environment, the sensor nodes are easily influenced by temperature, pressure, flow and other unexpected interferences. Some senor nodes may be dead because of energy exhaustion, internal device damage and so on. Suppose in some time slots, a sensor node (which may be the Nth node) did not send data to the cluster head because of being dead, then what the cluster head received is $y_{(N-1)} = (y_1, y_2, \ldots, y_{N-1})^T$. At this moment, the least squares that needs to be sent by the cluster head is

$$\hat{\theta} = \left(H_{N-1}^T H_{N-1}\right)^{-1} H_{N-1}^T y_{(N-1)}, \tag{3}$$

In formula (3), $H_{N-1} = (h_1, h_2, \ldots, h_{N-1})^T$.

It is not sensible for us to apply formula (3) calculating $\hat{\theta}$ directly, in other words, calculating $\left(H_{N-1}^T H_{N-1}\right)^{-1} H_{N-1}^T$ according to $h_i(i = 1, 2, \ldots, N-1)$. The least squares estimation expression involves the generalized inverse operation of matrix. If the generalized inverse operation of $H_{N-1}^T H_{N-1}$ is calculated directly, additional large data storage and computation are needed because of the numerous water environment parameters. It may even can't be calculated in real time in dynamic systems. Therefore, the study on data fusion algorithms which matching the condition of actual computing devices has important theoretical and practical significance.

4.2　The Forward and Backward Algorithms of Recursive Least Squares

Because of the relatively fixed storage and calculation, the low computation complexity and high stability, convenient implementation on computing devices, the recursive algorithm has been widely applied to the information processing and fusion field. And it is especially suitable for real time processing systems [17–19].

Considering the following equations

$$y = Hx, \tag{4}$$

where H is an $M \times d$ matrix and it is column full rank. The least squares can be used to calculate $x = (H^T H)^{-1} H^T y$.

If $H_M := H$ increases by one line, the result becomes

$$H_{M+1} = \begin{pmatrix} H_M \\ h_{M+1} \end{pmatrix}, \tag{5}$$

where h_{M+1} is the row vector.

- Matrix inversion lemma

$$(A + xy^T)^{-1} = A^{-1} - \frac{A^{-1}xy^T A^{-1}}{1 + y^T A^{-1}x}, \tag{6}$$

$$(A \pm B^T CB)^{-1} = A^{-1} \mp A^{-1}B^T (C^{-1} \pm BA^{-1}B^T)^{-1} BA^{-1}. \tag{7}$$

According to the above lemma given by [16], the forward and backward algorithms of recursive least squares can be deduced.

- Backward recursive algorithm

$$\begin{aligned}
(H_{M+1}^T H_{M+1})^{-1} &= (H_M^T H_M + h_{M+1} h_{M+1}^T)^{-1} \\
&= (H_M^T H_M)^{-1} - \frac{(H_M^T H_M)^{-1} h_{M+1} h_{M+1}^T (H_M^T H_M)^{-1}}{1 + h_{M+1}^T (H_M^T H_M)^{-1} h_{M+1}}
\end{aligned} \tag{8}$$

- Forward recursive algorithm

$$\begin{aligned}
(H_M^T H_M)^{-1} &= (H_M^T H_M + h_{M+1} h_{M+1}^T - h_{M+1} h_{M+1}^T)^{-1} \\
&= (H_{M+1}^T H_{M+1} - h_{M+1} h_{M+1}^T)^{-1} \\
&= (H_{M+1}^T H_{M+1})^{-1} + \frac{(H_{M+1}^T H_{M+1})^{-1} h_{M+1} h_{M+1}^T (H_{M+1}^T H_{M+1})^{-1}}{1 - h_{M+1}^T (H_{M+1}^T H_{M+1})^{-1} h_{M+1}}
\end{aligned} \tag{9}$$

If the increased H_{M+1} is not a row vector but a matrix, the forward and backward recursive algorithms of the matrix version can be obtained according to the second formula of Matrix inversion lemma (formula(7)).

4.3 Data Fusion Based on Forward Recursive Least Squares

In response to the problem stated above (in Sect. 4.1), the latest data fusion result of $N-1$ sensors can be easily obtained without calculating $(H_{N-1}^T H_{N-1})^{-1}$ by using the forward recursive least squares algorithm

$$(H_{N-1}^T H_{N-1})^{-1} = (H_N^T H_N)^{-1} + \frac{(H_N^T H_N)^{-1} h_N h_N^T (H_N^T H_N)^{-1}}{1 - h_N^T (H_N^T H_N)^{-1} h_N}. \tag{10}$$

The algorithm analyses are as follows:

1. During the running of LEACH, the cluster head should be re-selected after a certain period of time. After the cluster head is re-selected, the cluster head will send broadcast packets to the ordinary nodes. The ordinary node i chooses to join the nearest cluster and sends its own ID and the parameter h_i of measurement equipment to the cluster head. After collecting the IDs and parameters of all N ordinary nodes in the cluster, the time slot can be assigned to each ordinary node by the cluster head and H_N can be stored for later use. We only need to calculate $\left(H_N^T H_N\right)^{-1}$ once and store the result so as to obtain $\widehat{\theta}$ when data fusion is performed. For this reason, when the cluster head uses the forward recursive least squares algorithm to fuse data sent by the $N-1$ sensors, the generalized inverse operation of $H_{N-1}^T H_{N-1}$ is avoided. The computation complexity of the fusion algorithm can be reduced greatly.

2. If h_{ii} is the ith diagonal element of $H_N \left(H_N^T H_N\right)^{-1} H_N^T$, then $h_i \left(H_N^T H_N\right)^{-1} h_i^T = h_{ii}$. If we calculate each diagonal element h_{ii} beforehand and put it into formula (10), the calculations will be further simplified.

3. If two or more ordinary nodes do not send data to cluster head, $\widehat{\theta}$ can be calculated based on the matrix version of the forward recursive algorithm.

4.4 Backward Recursive Least Squares in the Stage of Cluster Head Rotation

In the stage of cluster head rotation, the measurement equipment parameter matrix H and the calculated $\left(H^T H\right)^{-1} H^T$ are stored. But usually the ordinary nodes connects into the channel through a competitive method, so the IDs of the ordinary nodes and the measurement equipment parameters are received in specified order by the cluster head. In fact, the cluster head can use the backward recursive least squares algorithm to compute $\left(H^T H\right)^{-1} H^T$ according to this characteristic (formula (8)). In this way, not only the computation complexity can be reduced but also the networking time can be shortened.

5 Conclusion and Future Work

By deploying lots of sensor nodes in the target area of rivers and lakes to construct a integrated water environment monitoring network, water environment changes and water pollution emergencies can be monitored automatically and continuously. Data fusion is used to reduce the amount of data sent and received, which can effectively save the energy of the whole network. Recursive algorithm can decrease the computation complexity of data fusion to a large extent.

For a cluster, the use of data fusion can reduce the energy consumption of sending data by the cluster head to $\frac{1}{N}$ with data fusion (N is the number of ordinary nodes in a cluster). If the entire network has n sensor nodes and the ratio of cluster head is p, then the use of data fusion will reduce the entire network energy consumption to $\frac{1}{n(1-p)}$ of the original. However, the increased energy consumption caused by data fusion is only some energy overhead of matrix calculations. $\left(H^T H\right)^{-1}$ and $\left(H^T H\right)^{-1} H^T$, which are used in the data fusion phase, have been calculated in advance, and some other matrix such as $\left(H_{N-1}^T H_{N-1}\right)^{-1}$ can be calculated using the recursive method, so the energy consumed by the matrix calculations is negligible compared to the saved in data fusion. The farther the cluster is, the more significant the result is.

In our future work, we will concentrate on uncertainty theories and multi-sensor data fusion. Furthermore, we will look for an intelligent and distributed data fusion algorithm and apply it in water environment monitoring.

Acknowledgements The authors acknowledge the financial support by the National Natural Science Foundation of China under Grant No. 61472343.

References

1. Liu, J.G., Yang, W.: Water sustainability for China and beyond. Science **337**(1), 649–650 (2012)
2. Uckelmann, D., Harrison, M., Michahelles, F.: The framework of IoT-Internet of Things Technology and its Impact on Society. Science Press, Beijing (2013)
3. Anderson, S.P., Bales, R.C., Duffy, C.J.: Critical zone observatories: building a network to advance interdisciplinary study of earth surface processes. Mineral. Mag. **72**(1), 7–10 (2008)
4. Kampe, T.U., Johnson, B.R., Kuester, M., et al.: NEON: the first continental-scale ecological observatory with airborne remote sensing of vegetation canopy biochemistry and structure. J. Appl. Remote Sens. **4**(1), 043510–043510-24 (2010)
5. Zacharias, S., Bogena, H., Samaniego, L., et al.: A network of terrestrial environmental observatories in Germany. Vadose Zone J. **10**(3), 955–973 (2011)
6. Li, X., Cheng, G.D., Liu, S., et al.: Heihe watershed allied telemetry experimental research (HiWATER): scientific objectives and experimental design. Bull. Am. Meteorol. Soc. **94**(8), 1145–1160 (2013)
7. Appriou, A.: Uncertainty Theories and Multisensor Data Fusion. Wiley (2014)
8. Luo, J.H., Wang, Z.J.: Multi-sensor Data Fusion and Sensor Management. Tsinghua University Press, Beijing (2015). (in Chinese)
9. Lee, J.: Optimal power allocating for correlated data fusion in decentralized WSNs using algorithms based on swarm intelligence. Wireless Netw. **23**(5), 1655–1667 (2017)
10. Yu, Y.: Consensus-based distributed mixture Kalman filter for maneuvering target tracking in wireless sensor networks. IEEE Trans. Veh. Technol. **65**(10), 8669–8681 (2016)
11. Smilde, A.K., Måge, I., Næs, T., et al.: Common and distinct components in data fusion. J. Chemom. **31**(7) (2017) (Version of Record online)

12. Paola, A.D., Ferraro, P., Gaglio, S.: An adaptive bayesian system for context-aware data fusion in smart environments. IEEE Trans. Mob. Comput. **16**(6), 1502–1515 (2017)
13. Lu, H., Li, Y., Chen, M., Kim, H., Serikawa, S.: Brain intelligence: go beyond artificial intelligence. Mob. Netw. Appl. 1–10 (2017)
14. Lu, H., Li, Y., Mu, S., Wang, D., Kim, H., Serikawa, S.: Motor anomaly detection for unmanned aerial vehicles using reinforcement learning. IEEE Internet of Things J. https://doi.org/10.1109/JIOT.2017.2737479
15. Serikawa, S., Lu, H.: Underwater image dehazing using joint trilateral filter. Comput. Electr. Eng. **40**(1), 41–50 (2014)
16. Wang, S.G., Shi, J.H., Yin, S.J., et al.: Introduction to Linear Models. Science Press, Beijing (2017). (in Chinese)
17. Wu, Z.S., Wang, Y.P.: Electromagnetic scattering for multilayered sphere: recursive algorithms. Radio Sci. **26**(6), 1393–1401 (2017)
18. Felis, M.L.: RBDL: an efficient rigid-body dynamics library using recursive algorithms. Auton. Robots **41**(2), 495–511 (2017)
19. Hong, X., Gao, J., Chen, S.: Zero-attracting recursive least squares algorithms. IEEE Trans. Veh. Technol. **66**(1), 213–221 (2017)

Modeling and Evaluating Workflow of Real-Time Positioning and Route Planning for ITS

Ping Liu, Rui Wang, Jie Ding and Xinchun Yin

Abstract Intelligent Traffic Systems (ITS), as integrated systems including control technologies, communication technologies, vehicle sensing and vehicle electronic technologies, have provided valuable solutions to the increasingly serious traffic problems. Hence, in order to achieve efficient management of all types of transportation resources and make better use of ITS, it is necessary and significant to continue study in depth on the architecture and performance of ITS. This paper adopts one kind of stochastic process algebra (SPA)—Performance Evaluation Process Algebra (PEPA) to model and evaluate the process of real-time positioning and route planning in ITS. Meanwhile, the fluid flow approximation is employed to conduct a performance analysis through PEPA models, then the maximize utilization and the throughput of the system can be achieved and analyzed.

Keywords Intelligent traffic system · PEPA · Performance evaluation

1 Introduction

Since 1999, benefiting from the revolution in communication and embedded computing technology, and the continuous improvement on a variety of sensors, mobile devices, GPS devices and actuators, Internet of things (IoT) has begun to develop rapidly. IoT has been regarded as the next generation of the Internet by many researchers. Furthermore, IoT has been successfully applied in many areas, such as: intelligent transportation, smart city, smart facility, smart farm and smart healthcare [1]. Through referring to the existing research results and combining the established architecture for IoT and ITS, this paper will model and generate performance evalu-

P. Liu · R. Wang · J. Ding(✉)
School of Information Engineering, Yangzhou University, Yangzhou, China
e-mail: jieding@yzu.edu.cn

J. Ding
State Key Laboratory for Novel Software Technology, Nanjing University, Nanjing, China

X. Yin
Guangling College, Yangzhou University, Yangzhou, China

© Springer International Publishing AG 2018
H. Lu and X. Xu (eds.), *Artificial Intelligence and Robotics*,
Studies in Computational Intelligence 752,
https://doi.org/10.1007/978-3-319-69877-9_30

ation for the working process of real-time positioning and route planning system in ITS.

Inspired by these references and the existed research results, we sum up a general workflow of ITS. And the four modules in Fig. 1 and the working process in Fig. 2 are also originated from the workflow of ITS. In addition, we divide the general workflow of ITS into four modules: vehicle terminals module (including vehicles and vehicles devices), wireless network module (such as Bluetooth, WiFi, 3G, 4G), GPS module (including positioning satellites and GPS system) and ITS management systems module.

As known to all, ITS is an integrated system including control technology, communication technology, vehicle sensing and vehicle electronics technologies and it also provides solutions to the increasingly serious traffic problems (such as traffic management, road safety). In addition to this, with the development of cloud computing technology, big data technology (image and data processing) [2, 3] and artificial intelligent technology [4], and ITS also linked to these technologies closely. Furthermore, with the deepening of research, ITS is divided into four main subsystems: Advanced traveler Information System (ATIS), Advanced Traffic Management System (ATMS), Advanced Public Transportation System (APTS) and Emergency Management System (EMS) [5]. Furthermore, in reference [6], authors introduce the existing Vehicular Cloud Computing (VCC) technology. The authors also design an architecture for VCC by using computing and storage devices in vehicles and have applied it to more vehicular communication systems.

In consideration of the property of working process in the real-time positioning and route planning system, we adopt a type of stochastic process algebra (SPA)— Performance Evaluation Process Algebra (PEPA) (see [7]) to model it. In order to evaluate the performance of the established system, fluid approximation method will be applied. In [8, 9], the authors express the discrete state space into continuous in fluid approximation. Furthermore, continuous time Markov chains (CTMC) that PEPA models on the strength of is converted to ordinary differential equations (ODEs) [10]. Therefore, by adopting the numerical solution of ODEs, we can easily acquire relative performance of the corresponding PEPA models (i.e. real-time positioning and route planning system).

Section 2 briefly introduces ITS and real-time positioning and route planning system. Section 3 first gives the corresponding system model of real-time positioning and route planning. Then, shows the performance evaluation and analysis of the system model. Section 4 concludes this paper.

2 ITS and Real-Time Positioning and Route Planning System

During the rapid development process of transport sector, many kinds of traffic problems become more and more visible. A growing number of problems in traffic management (such as vehicle management, transport facilities management and problem of road use) become an obstacle of the transport development in most countries. As an integrated system, ITS is made up of satellite positioning system, mobile vehicle terminals, wireless network and intelligent traffic management system. Meanwhile, ITS provides efficient and feasible solutions to these traffic management problems.

This paper mainly focuses on the global architecture of positioning and real-time traffic status query system in ITS. Through designing working process and conducting performance of the real-time positioning and route planning system, ITS can provide reliable traffic information and convenient travel route to users. Furthermore, ITS can also assist traffic management, improve the utilization of road and reduce traffic pressure. From reference [11], it is easy to know that an autonomic control loop consists of monitor module, analyze module, plan module and execute module. Meanwhile, in [11], the authors also provide and describe an autonomic system model for IoT. Furthermore, in reference [12], authors adopt the compact genetic algorithm for solving the carpool services problem and they also show us a fundamental framework of intelligent carpool system.

Inspired by references [11, 12] and according to the whole working process of the real-time positioning and route planning system, we divide the system into four modules (as shown in Fig. 1): the user module (UM), the information acquisition module (IAM), the information processing module (IPM) and the information storage module (ISM). The specific description of the four modules are listed as follows:

- The main function of UM is to send traffic query and analysis request and receive the corresponding response information.
- The IAM consists of two sub-modules: GPS and traffic acquire devices (TAD). Once receiving the corresponding requests, GPS utilizes satellites to obtain the location information and TAD uses all kinds of video image processing equipment acquire traffic information. Then, GPS and TAD send the acquired information to ISM through wireless network.
- The IPM is also composed of two sub-module. One sub-module is mainly responsible for communicating with other modules (i.e. receive requests and feedback

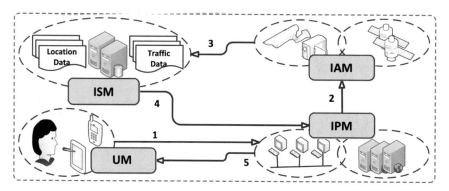

Fig. 1 Modules of real-time positioning and route planning system

responses). The other sub-module is mainly used to analyze the real-time traffic and users request, then provide reasonable path planning to users.

- The ISM is mainly used to store all kinds of information. This module reduces the complexity of information processing, improving the processing efficiency of the system and shortening the users query time largely.

3 Modelling and Performance Evaluation

3.1 PEPA Model of the System

This section is mainly to describe the overall working process of real-time positioning and route planning. From Fig. 2, it is easy to find that the whole process is divided into six components: User, GPS, traffic acquire devices (TAD), intelligent traffic system (ITS), analysis module (AM) and storage module (SM). Moreover, in Fig. 2, rectangles denote individual activities, round rectangles represent shared activities and diamonds denote choice. Meanwhile, it is easy to find three kinds of arrows are used in Fig. 2. Solid arrows are used to describe the execution sequences of different components. Dotted arrows represent the specific execution sequences of all activities within every component. Shorted dotted arrows denote a choice of execution sequences of some activities within SM module.

The whole specific workflow of the real-time positioning and route planning system is stated as follow:

1. User transmits a request to ITS for real-time positioning and route planning (i.e. route_ plan_req).
2. First of all, ITS will obtain position information of the destination upon receiving the User's request. In order to achieve this goal, ITS will then obtain the location of destination by delivering request to GPS (i.e. site_req).

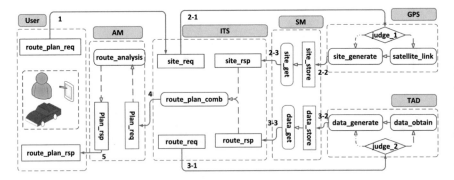

Fig. 2 Working process of real-time positioning and route planning system

3. After step 2, then, GPS will determine whether the corresponding satellite has been connected (i.e. *judge_1*). If not, next, GPS establishes a connection to the usable satellite (i.e. *satellite_link*) and generates the specific location information rely on the satellite positioning data (i.e. *site_generate*). Otherwise, GPS generates the position information directly.

4. In order to facilitate subsequent users to search the location information of the same area, GPS stores the generated position information into the storage module (i.e. *site_store*). During storage process of location information, the system will transform the data format which is from GPS. Simultaneously, SM obtains the transformed location information (i.e. *site_get*) and returns the information to ITS (i.e. *site_rsp*).

5. After step 4, ITS then issue a request to TAD to obtain route planning (i.e. *route_req*).

6. As soon as accepting the request from ITS, TAD will first judge whether the real-time traffic information has been obtained. If not, TAD first acquires traffic data (i.e. *data_obtain*) and then transforms the acquired data into real-time traffic information (i.e. *data_generate*). Otherwise, TAD generates real-time traffic information straightway.

7. Similar to 4, in order to facilitate the follow-up users to search real-time traffic information of the corresponding area, TAD saves the generated traffic data into SM (i.e. *data_store*). In the process of storing the traffic data, the system will transform the data format which is from TAD. Whereafter, SM gets the transformed traffic information (i.e. *data_get*) and feedbacks the information to ITS (i.e. *route_rsp*).

8. Once receiving information of location and real-time traffic, ITS will combine these information (i.e. *site_route_comb*) and request AM to obtain the real-time route planning to the destination (i.e. *plan_req*). Then, AM will make an intelligent analysis on the consolidated request from ITS (i.e. *route_analysis*) and feedback the specific route planning to ITS (i.e. *plan_rsp*). Later, ITS forwards the finally specific route planning to User (i.e. *route_plan_rsp*).

9. At last, after finishing the above steps, all modules: User, AM, ITS, SM, GPS and TAD will operate reset actions (i.e. reset 1, 2, 3, 4, 5 and 6) those we omit in Fig. 3.

Finally, we will give out the PEPA model of the component AM of real-time positioning and route planning system (the PEPA models of the other system components *User, ITS, GPS, SE* and *TAD*) are omitted here):

$$AM_1 \stackrel{\text{def}}{=} (plan\_req, r_{plan\_req}) \cdot AM_2$$

$$AM_2 \stackrel{\text{def}}{=} (route\_analysis, r_{route\_analysis}) \cdot AM_3$$

$$AM_3 \stackrel{\text{def}}{=} (plan\_rsp, r_{plan\_rsp}) \cdot AM_4$$

$$AM_4 \stackrel{\text{def}}{=} (reset5, r_{reset5}) \cdot AM_1$$

The system model is

$$(User[a] \underset{\{L_1\}}{\bowtie} ITS[b] \underset{\{L_2\}}{\bowtie} AM[c]) \underset{\{L_3\}}{\bowtie} (GPS[d] \underset{\{L_4\}}{\bowtie} SM[e] \underset{\{L_5\}}{\bowtie} TAD[f])$$

where

$L_1 = \{route\_plan\_req, route\_plan\_rsp\}$,
$L_2 = \{plan\_req, plan\_rsp\}$,
$L_3 = \{site\_req, site\_rsp, route\_req, route\_rsp\}$,
$L_4 = \{site\_store\}$,
$L_5 = \{data\_store\}$.

Here we list out two tables: Tables 1 and 2. Table 1 shows the duration and rate of all activities and Table 2 shows the specific number of all components. Furthermore, in Table 1, the duration represents the time required for the system to finish the corresponding activity and the rate represents the number of times that system finishes the corresponding activity within one second. And the most parameters in Table 1 are acquired by debugging on practical environment of Internet. Because of

Table 1 Duration and rate of all activities

Action	Duration (s)	Rate
route_plan_req	0.00037	2702.70
site_req	0.00048	2083.33
judge_1	0.001	1000
satellite_link	1.5	0.67
site_generate	0.67	1.49
site_store	1.081	0.93
site_get	0.713	1.40
site_rsp	0.00033	3030.30
route_req	0.00035	2857.14
judge_2	0.001	1000
data_obtain	0.67027	1.49
data_generate	0.70	1.43
data_store	1.081	0.93
data_get	0.713	1.40
route_rsp	0.00056	1785.71
site_route_comb	0.0025	400
plan_req	0.00035	2857.14
route_analysis	2.8	0.36
plan_rsp	0.03	33.33
route_plan_rsp	0.00351	284.90
reset1	1.0	1
reset2	0.0001	10000
reset3	0.001	1000
reset4	0.0001	10000
reset5	0.0001	10000
reset6	0.0001	10000

Table 2 Number of components

Component	Number	Component	Number
User	300	GPS	60
TAD	60	ITS	80
AM	100	SM	100

equipment limitation, it is difficult to acquire the specific data of all components. As a result, the specific number of all components in Table 2 are supposed according to Internet resources and practical situation.

3.2 Performance Analysis

The specific dynamic performance of the real-time positioning and route planning is presented in this section (Figs. 3, 4 and 5). As shown in these figures, response time of the whole process of the model, throughput of *data_obtain*, *route_analysis*, *data_get* and *site_route_comb* will be respectively analyzed.

First of all, we take the number of Users as constant which is presented in Table 2. Next, we make an analysis on the probability of Users finish the whole process of real-time route planning. So, in Fig. 3, it is easy to find that when the number of Users increases, the probability of Users complete the route planning query is lower.

The throughput of *data_obtain*, *route_analysis*, *data_get* and *site_ route_ comb* are analyzed in Figs. 4 and 5. From the two figures, it is easy to find that when the requests from Users increase, the throughput of these activities becomes larger too. However, the curve will become smoothly when throughput approaches the maximum of system. That means, the whole system has become saturated and

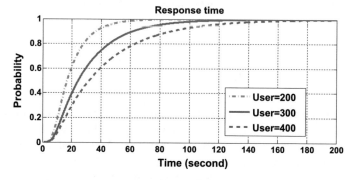

Fig. 3 Response time of users complete the process of real-time route planning

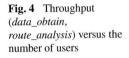

Fig. 4 Throughput
(*data_obtain*,
route_analysis) versus the
number of users

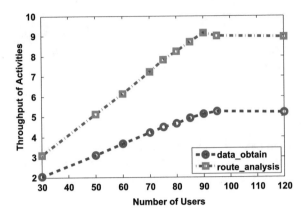

Fig. 5 Throughput
(*data_get*, *site_route_comb*)
versus the number of users

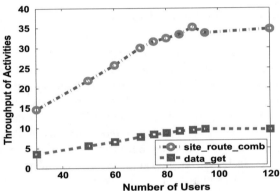

every components has also reached the highest processing speed of accepting request and feedback response under steady state of the system.

Through analyzing the response time and throughput, it is helpful to test the ability of the system to process Users' requests in practice and maximize the utilization of the system.

4   Conclusion and Future Work

This paper employs formalization PEPA to model the whole workflow of real-time positioning and route planning in ITS. As in practice, an efficient ITS is of great significance for the management of traffic information and traffic infrastructure. At the same time, performance evaluation of the system is helpful for supervisor maximize the use of the system. In our future work, we will concentrate on intelligent decision and intelligent push of real-time traffic. Furthermore, we will look for a rational and effective intelligent decision-making algorithm and apply it in ITS.

Acknowledgements The authors acknowledge the financial support by the National NSF of China under Grant No. 61472343.

References

1. Wen, Z., Liu, X., Xu, Y.: A RESTful framework for internet of things based on software defined network in modern manufacturing, **84**(1), 361–369 (2016)
2. Li, Y., Lu, H., Li, J., Li, X., Li, Y., Seiichi, S.: Underwater image de-scattering and classification by deep neural network. Comput. Electr. Eng. **54**, 68–77 (2016)
3. Xing, X., He, L., Shimada, A., Taniguchi, R.-i., Lu, H.: Learning unified binary codes for cross-modal retrieval via latent semantic hashing. Neurocomputing **213**, 191–203 (2016)
4. Lu, H., Li, Y., Chen, M., Kim, H., Serikawa, S.: Brain Intelligence: Go Beyond Artificial Intelligence. Mobile Networks and Application, pp. 1–10 (2017)
5. Singh, B., Gupta, A.: Recent trends in intelligent transportation systems: a review, **9**, 30–34 (2015)
6. Whaiduzzamana, Md., Sookhaka, M., Gania, A., Buyyab, R.: A survey on vehicular cloud computing, **40**, 325–344 (2014)
7. Hillston, J.: A Compositional Approach to Performance Modelling. Cambridge University Press, Cambridge (1996)
8. Hillston, J.: Fluid flow approximation of PEPA models. In: Processing of the second International Conference on the Quantitative Evaluation of System, IEEE Computer Society, pp. 33–43 (2005)
9. Ding, J.: A comparison of fluid approximation and stochastic simulation for evluating content adaptation systems. Wirel. Pers. Commun. **84**(1), 231–250 (2015)
10. Ding, J., Hillston, J.: Numerically representing stochastic process algebra models. Comput. J. **55**(11), 1383–1397 (2012)
11. Ashraf, Q.M., Habaebi, M.H., Islam, Md.R.: TOPSIS-based service arbitration for autonomic internet of things, **4**, 1313–1320 (2016)
12. Jiau, M.-K., Huang, S.-C.: Services-oriented computing using the compact genetic algorithm for solving the carpool services problem, **16**(5), 2711–2722 (2015)

Cost-Sensitive Collaborative Representation Based Classification via Probability Estimation Addressing the Class Imbalance Problem

Zhenbing Liu, Chao Ma, Chunyang Gao, Huihua Yang, Tao Xu, Rushi Lan and Xiaonan Luo

Abstract Collaborative representation has been successfully used in pattern recognition and machine learning. However, most existing collaborative representation classification methods are to achieve the highest classification accuracy, assuming the same losses for different misclassifications. This assumption may be ineffective in many real-word applications, as misclassification of different types could lead to different losses. Meanwhile, the class distribution of data is highly imbalanced in real-world applications. To address these problems, Cost-sensitive Collaborative Representation based Classification via Probability Estimation Addressing the Class Imbalance Problem method was proposed. The class label of test samples was predict by minimizing the misclassification losses which are obtained via computing the posterior probabilities. In this paper, a Gaussian function was defined as a probability distribution of collaborative representation coefficient vector and it was transformed into collaborative representation framework via logarithmic operator. The experiments on UCI and YaleB databases show that our method performs competitively compared with other methods.

Keywords Collaborative representation · Cost-sensitive learning Probability estimate · Loss function

Z. Liu · H. Yang · R. Lan (✉) · X. Luo
Guangxi Colleges and Universities Key Laboratory of Intelligent Processing of Computer Images and Graphics, Guilin University of Electronic Technology, Guilin, China
e-mail: rslan2016@163.com

C. Ma · C. Gao · T. Xu
School of Electronic Engineering and Automation, Guilin University of Electronic Technology, Guilin, China

C. Ma
Guangxi Colleges and Universities Key Laboratory of Intelligent Processing of Computer Images and Graphics, Guilin University of Electronic Technology, Guilin, China

H. Yang
School of Automation, Beijing University of Posts and Telecommunications, Beijing, China

© Springer International Publishing AG 2018
H. Lu and X. Xu (eds.), *Artificial Intelligence and Robotics*,
Studies in Computational Intelligence 752,
https://doi.org/10.1007/978-3-319-69877-9_31

1   Introduction

In recent year, cost-sensitive learning has been studied widely and become one of the most important topics for solving the class imbalance problem [1]. In [2], Zhou et al. studied empirically the effect of sampling and threshold-moving in training cost-sensitive neural networks, and revealed threshold-moving and soft-ensemble are relatively good choices in training cost-sensitive neural networks. In [3], Sun et al. proposed a cost-sensitive boosting algorithms, which are developed by introducing cost items into the learning framework of AdaBoost. In [4], Jiang et al. proposed a novel Minority Cloning Technique (MCT) for class-imbalanced cost-sensitive learning. MCT alters the class distribution of training data by cloning each minority class instance according to the similarity between it and the mode of the minority class. In [5], a new cost-sensitive metric was proposed by George to find the optimal tradeoff between the two most critical performance measures of a classification task-accuracy and cost. Generally, users focus more on the minority class and consider the cost of misclassifying a minority class to be more expensive. In our study, we adopt the same strategy to addressing this problem.

Motivated by probabilistic collaborative representation based approach for pattern classification [6] and Zhang's work [7], in this paper we propose a new method to handle misclassification cost and class-imbalance problem called Cost-sensitive Collaborative Representation based Classification via Probability Estimation Addressing the Class Imbalance Problem (CSCRC). In Zhang's cost-sensitive learning framework, posterior probabilities of a testing sample are estimated by KLR or KNN method. In [6], Probabilistic Collaborative Representation based approach for pattern Classification (ProCRC) is designed to achieve the lowest recognition errors and assume the same losses for different types of misclassifications, it is difficult to resolve the class imbalance problem. For this case, we introduce cost-sensitive learning framework into ProCRC, which not only derive the relationship between Gaussian function and collaborative representation but also resolve the cost-sensitive problem in [6]. Firstly, we use the probabilistic collaborative representation framework to estimate the posterior probabilities. The posterior probabilities are generated directly from the coding coefficients by using a Gaussian function and applying the logarithmic operator to the probabilistic collaborative representation framework, this explained clearly the l_2-norm regularized representation scheme used in collaborative representation based classifier (CRC). Secondly, calculating all the misclassification losses use Zhang's cost-sensitive learning framework. At last, the test sample is assigned to the class whose loss is minimal. Experimental results on UCI databases validate the effectiveness and efficiency of our methods.

2   Proposed Approach

In Cai's work [6], different data points x have different probabilities of $l(x) \in l_X$, where l(x) means the label of x, l_X means the label set of all candidate classes in X, and $P(l(x) \in l_X)$ should be higher if the l_2-norm of α is smaller, vice versa. One intuitive choice is to use a Gaussian function to define such a probability:

$$P(l(x) \in l_X) \propto exp\left(-c\|\alpha\|_2^2\right) \tag{1}$$

where c is a constant and data points are assigned different probabilities based on α, where all the data points are inside the subspace spanned by all samples in X. For a sample y outside the subspace, the probability as:

$$P(l(y) \in l_X) = P(l(y) = l(x)|l(x) \in l_X)P(l(x) \in l_X) \tag{2}$$

$P(l(x) \in l_X)$ has been defined in Eq. (7). $P(l(y) = l(x)|l(x) \in l_X)$ can be measured by the similarity between x and y. Here we adopt the Gaussian kernel to define it:

$$P(l(y) = l(x)|l(x) \in l_X) \propto exp(-k\|y - x\|_2^2) \tag{3}$$

where k is a constant, with Eqs. (1)–(3), we have

$$P(l(y) \in l_X) \propto exp(-(k\|y - X\alpha\|_2^2 + c\|\alpha\|_2^2)) \tag{4}$$

In order to maximize the probability, we can apply the logarithmic operator to Eq. (4). There is:

$$\begin{aligned} maxP(l(y) \in l_X) &= \max \ln(P(l(y) \in l_X)) \\ &= min_\alpha k\|y - X\alpha\|_2^2 + c\|\alpha\|_2^2 \\ &= min_\alpha \|y - X\alpha\|_2^2 + \lambda\|\alpha\|_2^2 \end{aligned} \tag{5}$$

where $\lambda = c/k$. Interestingly, Eq. (5) shares the same formulation of the representation formula of CRC [4], but it has a clear probabilistic interpretation.

A sample x inside the subspace can be collaboratively represented as: $x = X\alpha = \sum_{k=1}^{K} X_k \alpha_k$, where $\alpha = [\alpha_1; \alpha_2; \ldots; \alpha_k]$ and α_k is the coding vector associated with X_k. Note that $x_k = X_k \alpha_k$ is a data point falling into the subspace of class k. Then, we have

$$P(l(x) = k|l(x) \in l_X) \propto exp(-\delta\|x - X_k \alpha_k\|_2^2) \tag{6}$$

where δ is a constant. For a query sample y, we can compute the probability that $l(y) = k$ as:

$$P(l(y) = k)$$
$$= P(l(y) = l(x) | l(x) = k) \cdot P(l(x) = k) \tag{7}$$
$$= P(l(y) = l(x) | l(x) = k) \cdot P(l(x) = k | l(x) \in l_X) \cdot P(l(x) \in l_X)$$

Since the probability definition in Eq. (3) is independent of k as long as $k \in l_X$, we have $P(l(y) = l(x) | l(x) = k) = P(l(y) = l(x) | l(x) \in l_X)$. With Eqs. (5)–(7), we have

$$P(l(y) = k) = P(l(y) \in l_X) \cdot P(l(x) = k | l(x) \in l_X)$$
$$\propto exp(-\|y - X\alpha\|_2^2 + \lambda\|\alpha\|_2^2 + \gamma\|X\alpha - X_k\alpha_k\|_2^2) \tag{8}$$

where $\gamma = \delta/k$. Applying the logarithmic operator to Eq. (8) and ignoring the constant term, we have:

$$(\hat{\alpha}) = \arg\min_{\alpha}\{\|y - X\alpha\|_2^2 + c\|\alpha\|_2^2 + \|X\alpha - X_k\alpha_k\|_2^2\} \tag{9}$$

Refer to Eq. (9), let X_k' be a matrix which has the same size as X, while only the samples of X_k will be assigned to X_k' at their corresponding locations in X, i.e., $X_k' = [0, \ldots, X_k, \ldots, 0]$. Let $\bar{X}_k' = X - X_k'$. We can then compute the following projection matrix offline:

$$T = (X^T X + (\bar{X}_k')^T \bar{X}_k' + \lambda I)^{-1} X^T \tag{10}$$

where I denotes the identity matrix. Then, $\hat{\alpha} = Ty$.

With the model in Eq. (9), a solution vector $\hat{\alpha}$ is obtained. The probability P(l (y) = k) can be computed by:

$$P(l(y) = k) \propto exp(-(\|y - X\hat{\alpha}\|_2^2 + \lambda\|\hat{\alpha}\|_2^2 + \|X\hat{\alpha} - X_k\hat{\alpha}_k\|_2^2)) \tag{11}$$

Note that $\left(\|y - X\hat{\alpha}\|_2^2 + \lambda\|\hat{\alpha}\|_2^2\right)$ is the same for all classes, and thus we can omit it in computing P(l(y) = k). Then we have

$$P_k = exp\left(-\left(\|X\hat{\alpha} - X_k\hat{\alpha}_k\|_2^2\right)\right) \tag{12}$$

In cost-sensitive learning, the loss function is regarded as an objective function to identify the label of a test sample. In binary classification problem, there are two misclassification costs, and we denote the cost that misclassify positive class as negative class by C_{10}, and the cost by C_{01} conversely. Then a cost matrix can be constructed as shown in Table 1, where G_1, G_0 represents the label of minority class and majority class, respectively.

Table 1 The classification accuracy for the 5 methods on 10 data sets

Accuracy	CRC	SRC	SVM	ProCRC	CSCRC
Letter	0.9932	0.9945	0.9690	0.9653	0.9950
Balance	0.9829	0.9138	0.9786	0.9357	0.9919
Abalone	0.9147	0.9984	0.9982	0.9994	0.9997
Car	0.9982	0.9781	0.9995	0.9902	1.0000
Pima	0.9395	0.9779	0.8665	0.9053	0.9377
Nursery	0.9994	0.9356	0.9718	0.9947	0.9998
Cmc	0.9729	0.9900	0.9981	0.9994	0.9997
Haberman	0.9088	0.9905	0.9506	0.9478	0.9997
Housing	0.9934	1.0000	0.6750	0.9888	0.9919
Ionosphere	0.9794	0.6738	0.9913	0.9581	0.9716
Average	0.96824	0.94526	0.93986	0.96847	0.98870

It is well known that the loss function can be related to the posterior probability $P(\phi(y)|y) \approx P(l(y)=k)$. Then the loss function can be rewritten as follow:

$$loss(y, \phi(y)) = \begin{cases} \sum_{i=G_1} P_i C_{10} \text{ if } \phi(y) = G_0 \\ \sum_{i=G_0} P_i C_{01} \text{ if } \phi(y) = G_1 \end{cases} \tag{13}$$

The test sample y belongs to the class with higher probability. We can obtain the label of test sample y by minimizing Eq. (13):

$$L(y) = \arg \min_{i \in \{0, 1\}} loss(y, \phi(y)) \tag{14}$$

3 Results

Experiment 1 We compare the performance of these 5 methods (sparse representation based classification (SRC), CRC, SVM, ProCRC, CSCRC) on 10 UCI data sets, and the results are summarized in Tables 1 and 2. The last row of Table 1 is the average Accuracy value for the method on ten data sets. We select 31 positive samples and 31 negative samples randomly from data sets Haberman, Housing, Ionosphere and Balance as test samples, 41 positive samples and 41 negative samples as training samples; 61 positive samples and 61 negative samples as test samples, 101 positive samples and 101 negative samples as training samples from the other 6 data sets. The cost ratio (the cost of false acceptance respect to false rejection) set as 10. We perform the process for 50 times and get the average results.

On Letter, Balance, Abalone, Car, Nursery, Cmc and Haberman, our method achieves very high Accuracy value respect to the other four methods. One of the three data sets does not get the highest value of Accuracy, but we achieve the

Table 2 The average cost for the 5 methods on 10 data sets

Average cost	CRC	SRC	SVM	ProCRC	CSCRC
Letter	0.0663	0.0548	0.1616	0.2118	0.0050
Balance	0.0943	0.8619	0.0343	0.0643	0.0081
Abalone	0.3335	0.0016	0.0018	0.0050	0.0003
Car	0.0032	0.2179	0.0034	0.0737	0.0000
Pima	0.1694	0.1658	0.8173	0.5084	0.0623
Nursery	0.0065	0.6435	0.2416	0.0271	0.0002
Cmc	0.2056	0.1000	0.0048	0.0006	0.0003
Haberman	0.5075	0.0952	0.2941	0.2772	0.0003
Housing	0.0516	0.0000	2.0828	0.0113	0.0081
Ionosphere	0.1191	2.5719	0.0313	0.3906	0.1044
Average	0.1557	0.4713	0.3673	0.1570	0.0189

highest value of average Accuracy. The values of accuracy are higher than 0.93. In other words, our method have better performance than SRC, CRC, SVM, ProCRC and CSCRC.

We calculate the misclassification cost of this 5 method on 10 UCI data sets and summarized as Table 2. On Letter, Balance, Abalone, Car, Pima, Nursery, Cmc and Haberman, our method achieves very low average misclassification cost. In Table 1, SRC has the highest value on Pima, but CSCRC has the highest value of Average Cost on Pima. Obviously, CSCRC classify the positive samples correctly. Furthermore, the value of Accuracy and is lower than CRC on Housing and Ionosphere, but the value of Average Cost is inverse.

Experiment 2 Similarly, we compare the performance of these 5 methods (SRC, CRC, SVM, ProCRC, CSCRC) on Letter, and evaluate the performance via G-mean and Average Cost for the class-imbalance problem. In this experiment, we taken the imbalance ratio from [1, 2, ..., 10], respectively. The size of minority class is 30 and the majority class is 30 multiply the imbalance ratios in train set, accordingly. We select 61 positive samples and 61 negative samples as test set. The cost is set as mentioned above.

Note that there are also situations in which CSCRC is preferred. From the results on Figs. 1 and 2 we can see that CSCRC has higher G-mean than the other four methods except the imbalance ratio is 1. Meanwhile, CSCRC achieves the lowest Average Cost respect to the other methods. This suggests that Cascade can focus on more useful data. With the increasing of imbalance ratio, we have more training samples, and the proposed method can classify the samples correctly when the imbalance ratio is up to 4. Generally speaking, class-imbalance does affect the proposed method CSCRC. Concretely, CSCRC is not influenced by the distribution of samples, we can also get a better classify result when the imbalance ratio is high.

Fig. 1 The result of G-mean
on letter

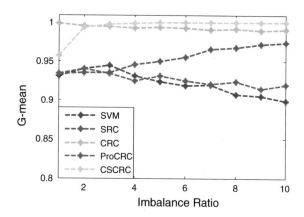

Fig. 2 The result of average
cost on letter

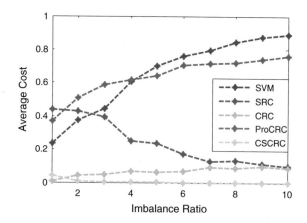

4 Conclusions

The class imbalanced datasets occurs in many real-world applications where the
class distributions of data are highly imbalanced. This paper, we propose a novel
method to handle misclassification cost and class imbalance problem called
Cost-sensitive Collaborative Representation Classification based Probability Estimation. The proposed approach adopted probabilistic model and sparse representation coefficient matrix to estimate the prior probability and then obtained the label
of a testing sample by minimizing the misclassification losses. The experimental
results show that the proposed CSCRC has a comparable or even lower average
cost with higher accuracy compare to the other four classification algorithm.

In order to simplify the cost matrix, we restrict our discussion to two-class
problems. So, extending our current work to multi-class scenario is a main research
direction for our future work.

Acknowledgements The authors want to thank the anonymous reviewers and the associate editor for helpful comments and suggestions. This work is supported by the National Natural Science Foundation of China (Grant Nos. 61562013, 21365008 and 61320106008), Guangxi Colleges and Universities Key Laboratory of Intelligent Processing of Computer Images and Graphics (No. LD16096x), the Center for Collaborative Innovation in the Technology of IOT and the Industrialization (WLW20060610).

References

1. Lan, R., Yang J.: Orthogonal projection transform with application to shape description. In: 2010 IEEE International Conference on Image Processing (2010)
2. Zhou, Z.H., Liu, X.Y.: Training cost-sensitive neural networks with methods addressing the class imbalance problem. IEEE Trans. Knowl. Data Eng. **18**(1), 63–77 (2006)
3. Sun, Y., Kamel, M.S., Wong, A.K.C., et al.: Cost-sensitive boosting for classification of imbalanced data. Pattern Recogn. **40**(12), 3358–3378 (2007)
4. Jiang, L., Qiu, C., Li, C.: A novel minority cloning technique for cost-sensitive learning. Int. J. Pattern Recogn. Artif. Intell. **29**(4) (2015)
5. George, N.I., Lu, T.P., Chang, C.W.: Cost-sensitive performance metric for comparing multiple ordinal classifiers. Artif. Intell. Res. **5**(1), p135 (2016)
6. Cai, S., Zhang, L., Zuo, W., et al.: A probabilistic collaborative representation based approach for pattern classification. Comput. Vis. Pattern Recogn. (2016)
7. Yin, Z., Zhi-Hua, Z.: Cost-sensitive face recognition. IEEE Trans. Pattern Anal. Mach. Intell. **32**(10), 1758–1769 (2010)

Motor Anomaly Detection for Aerial Unmanned Vehicles Using Temperature Sensor

Yujie Li, Huimin Lu, Keita Kihara, Jože Guna and Seiichi Serikawa

Abstract Aerial unmanned vehicle is widely used in many fields, such as weather observation, framing, inspection of infrastructure, monitoring of disaster areas. However, the current aerial unmanned vehicle is difficult to avoid falling in the case of failure. The purpose of this article is to develop an anomaly detection system, which prevents the motor from being used under abnormal temperature conditions, so as to prevent safety flight of the aerial unmanned vehicle. In the anomaly detection system, temperature information of the motor is obtained by DS18B20 sensors. Then, the reinforcement learning, a type of machine learning, is used to determine the temperature is abnormal or not by Raspberrypi processing unit. We also build an user interface to open the screen of Raspberrypi on laptop for observation. In the experiments, the effectiveness of the proposed system to stop the operation state of drone when abnormality exceeds the automatically learned motor temperature. The experimental results demonstrate that the proposed system is possibility for unmanned flight safely by controlling drone from information obtained by attaching temperature sensors.

Keywords Aerial unmanned vehicle · Anomaly detection · Temperature sensor

1 Introduction

Unmanned aerial vehicles flying through the air is unattended and controlled remotely. Drone, one kind of unmanned aerial vehicles, is widely used, such as, weather observation, spraying of agricultural chemicals, inspection of infrastruc-

Y. Li
Yangzhou University, 127 Huayang Rd., Hanjiang, Yangzhou 225127, China

H. Lu (✉) · K. Kihara · S. Serikawa
Kyushu Institute of Technology, 1-1 Sensui, Tobata, Kitakyushu 8048550, Japan
e-mail: dr.huimin.lu@ieee.org

J. Guna
University of Ljubljana, 1000 Ljubljana, Slovenia

© Springer International Publishing AG 2018
H. Lu and X. Xu (eds.), *Artificial Intelligence and Robotics*,
Studies in Computational Intelligence 752,
https://doi.org/10.1007/978-3-319-69877-9_32

ture, monitoring of disaster areas. Rescue drones, whose functions with capturing live footage of the disaster area, providing first aid kits and enabling Wi-Fi to the local area. However, the usage of drones in extreme environment causes many issues. It is reported that 418 accidents have been happened in USA. The Aeronautical Law, which is announced by the Japanese government shows that it is necessary to safety flying when using the drone over the area where people or houses are densely populated. To the best of our knowledge, there is no system concerning drone's motor abnormality detection yet. However, it is considered to be a very useful and important issue. Hence, in this paper, we develop a detection and protection system for landing the drone when it reaches the abnormal temperature.

One kind of the accidents of drone is the fall accidents, which includes short flight time due to shortage of the capacity of battery, malfunction in motor and DC errors, loss of communication etc. In order to solve these problems, it is necessary to design a new anomaly detection system. In this paper, we consider the motor abnormal when operating. One the one hand, the effect of the design life of the motor is the bearing life. The bearing life can be expressed by the following two kinds. The first one is the grease life where grease deteriorates due to heat. The second one is the mechanical life due to rolling fatigue. In most cases, the grease life is dominant in motors, because it has a greater influence on grease life due to heat generation than the influence of mechanical life due to the load applied to the bearings. The most important factor affects the service life of the bearing grease is temperature, and its lifetime mainly depends on the temperature. Consequently, we use the temperature of the motor to detect abnormality of the motor of the drone. That is, the failure of the motor is usually reflected the temperature anomalies.

We propose a motor anomaly detection and falling protection system using temperature sensors and ARM Cortex-A53 CPU (Raspberrypi 2B) for the purpose of controlling the usage of abnormal temperature of motor. Temperature sensors are adhered to the motor unit of the drone. The abnormal temperature of the motors is decided on the Raspberrypi, which is set in the drone. The drone is automatically landing if the temperature is abnormal, and it is taken off again if the temperature is normal. With this system, there is no fear of falling down when using the drone in extreme environments where failures are not permitted. In addition, this automatic motor anomaly detection system increases safety, leading to more active use of automatic piloting, and consequently decreasing fall that is caused by communication loss and the errors by controller steering.

The organization of this paper in the following is: In Sect. 2, some related works about motor anomaly detection are reviewed. Section 3 shows the system configuration of the proposed system. Experimental results are shown in Sect. 4. Finally, we conclude this paper in Sect. 5.

2 Related Works

There are many methods for detecting anomaly of the motor. In this section, we summarize all of these methods and discuss the benefits and disadvantages of these methods.

Firstly, there is an inspection system by abnormal sound detection [1]. It is a method to use the abnormal sound inspection in case-based identification algorithm. In this method, it is necessary to calculate the voice data by feature vector beforehand and calculate it based on the establishment distribution model, which is a kind of supervised learning. Supervised learning requires a training dataset of sample elements along with the desired classification of each of these samples. However, in most cases, it is difficult to training the datasets in extreme environments. So, wrong detection is caused using sound inspection in high noisy background (Fig. 1).

Secondly, there is a system in which a motor is photographed and an alarm is output when the detected temperature becomes equal to or higher than the threshold temperature in the infrared image [2, 3]. This system has a problem that it is difficult to set a thermal imaging camera around the drone. It is difficult to both install a camera to capture a small infrared image and determine the temperature change through low quality infrared images (Fig. 2).

Thirdly, the fan sensor constantly monitors the rotation speed of the fan motor and outputs an alarm signal when abnormal rotation occurs [4, 5]. It is difficult to judge whether the reduction of the rotation speed is due to abnormality or low speed

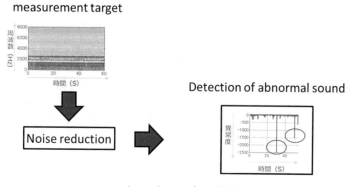

Operating sound collected from the measurement target

Detection of abnormal sound

Noise reduction

Learning and modeling operating sound at normal time
· Determine "probability of abnormality" from the degree of divergence from the normal model of the acquired operation sound
· Issue an alert according to preset threshold

Fig. 1 Inspection system using abnormal sound detection

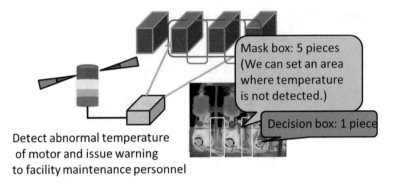

Fig. 2 Thermal imaging based abnormal temperature detection system

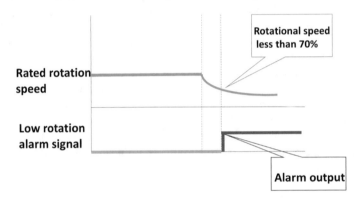

Fig. 3 Outline drawing of fan sensor system

in the drone that changes speed. Furthermore, attaching a fan sensor on the motors will affect the flying of drone (Fig. 3).

As described in above, the conventional anomaly detection methods have the disadvantages in detecting abnormality of the droning motors. In this paper, it was made possible to develop an anomaly detection system on a drone which is difficult with these methods using a temperature sensor (DS18B20), and Raspberrypi 2B.

3 System Configuration

In this paper, as a system to prevent the abnormality falling of drone under temperature condition for the motor. We build a system to landing when the motor temperature exceeds a certain temperature.

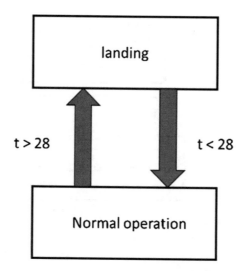

Fig. 4 System flowchart

If the temperature of the motor exceeds a certain threshold value as an abnormal condition, it is defined as abnormal. Previous work has showed that the abnormal temperature of drone is between 70–80 °C. In addition, in this study, because the drone's flying time is as short as 10 min, 28 °C is adopted as a substitute threshold.

3.1 System Overview

Figure 4 shows the flowchart of the system in this system. The model proposed in this research consists of two devices, a drone real machine equipped with what is necessary for experiments and a hardware for observation. The drone side is equipped with a temperature sensor, Raspberrypi. Hardware for observation use laptop. A block diagram of the system is shown in Fig. 5.

The processing sequence of the system is

1. Make Raspberrypi operable on a laptop
2. From takeover command from Raspberrypi to drone
3. Drone advance after the takeoff
4. Measure the temperature of the motor part during flight
5. Automatically landing when exceeding the threshold set as abnormality (28°).

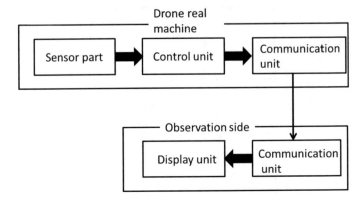

Fig. 5 System working diagram

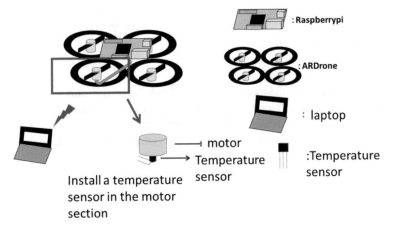

Fig. 6 Conceptual diagram of system

3.2 Around Drones

3.2.1 Drone Composition

The structure of the drone is made up of four units: a detection unit, an actual machine (drone), a control unit, and a communication unit. A temperature sensor is used for the detection unit and Raspberrypi is used as the control part. As a control method, we implement our own python program on Raspberrypi and operate the drone's movement automatically. This program is described in detail in Sect. 3.2.3. The communication section implements VNC and enables communication between Raspberrypi and the laptop. The configuration is shown in the Fig. 6.

Fig. 7 Circuit diagram around the sensor

3.2.2 Detection Unit

A temperature measurement circuit using a temperature sensor is used for the detection section. The measurement circuit is shown in the Fig. 7. In this paper, a temperature sensor is installed as shown in Fig. 8 on the part where the motor is exposed (lower side of the motor case) to measure the temperature of the motor. As 1-wire has an A/D converter build in the sensor itself, there is no signal deterioration due to wiring length and temperature data is converted to digital inside the sensor, so there is no error in signal transmission. In RaspbianOS, this sensor is used this time from the point that it can correspond to a 1-wire device by just loading a module for 1-wire.

3.2.3 Control Unit

The control unit uses Raspberrypi. Raspberrypi used this time is modelB+. This is developed as a single-board computer for education which is cheaper than the Raspberrypi Foundation. It has the feature that it can be used by installing power supply and SD card storage instead of mounting the internal hard disk etc. This time, modelB+ has selected a function from a cheap price with a single function attached.

There are two things that the control section did.

1. Construction of an environment to control from Raspberrypi to drone.
2. Based on the information obtained from the temperature sensor, let it process whether the drones should land.

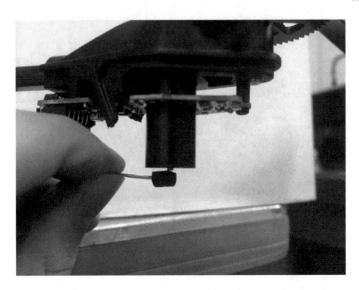

Fig. 8 Actual machine drawing (motor part)

As a connection from Raspberrypi to the drone, make a Wi-Fi connection with the drones as parent devices. The SSID of the drone used this time is AR.Drone 2-132506, IP address is 192.168.1.1.

And python is used as a programming language for controlling drone. Then, measure the temperature of the motor part of the drone and measure the temperature exceeding 28 °C set as the threshold value of the brushless motor. Implement the landing program. The flowchart of the system is shown in the figure. The program summarizing the whole is shown in Fig. 9.

As the flow of the landing system when measuring 28 °C which is the threshold judged to be abnormal in the brushless motor in this study, firstly, we took off the drone with instructions from Raspberrypi. At the stage of takeoff, temperature measurement is started with a temperature sensor attached to the motor. Since the result of temperature measurement is assigned to temp_c is displayed each time measurement is performed. Measurement of the temperature to judge whether the motor temperature is abnormal is carried out every 5 s, the measurement result is transmitted from the temperature sensor to Raspberrypi, and when the measurement result exceeds 28 °C set as the threshold value, the system detects the abnormal and land automatically.

After landing, if temperature measurement continues and the measurement result falls below 28 °C, the flight from takeoff is started again. If the measurement result does not exceed 28 °C, it is judged that the flight is possible, and after 5 s, the forward flight continues until the measurement is made again. In practice, the speed of the drone can be set in the range of 0–1 on the program. The speed is set to 0.05 because ease of measurement is required and there was no necessity of high speed operation.

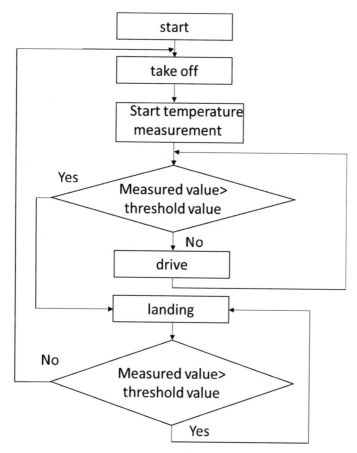

Fig. 9 System flowchart

4 Simulation Results

In this paper, a system was constructed to achieve the purpose of landing when the motor temperature exceeded a certain value. Therefore, in the simulation, we confirmed whether this objective was achieved. When the system was operated with the threshold set at 28 °C. It was confirmed that the system on Raspberrypi was used to detect the abnormal and the drone was lowered when the motor temperature reached 28 °C after the start of flight.

5 Conclusion

In this paper, we have proposed a motor anomaly detection system for unmanned vehicles using temperature sensors. We were able to construct a system to deal with motor abnormality, which is one of the causes of drone falling, using a temperature sensor and Raspberrypi. In the simulation, when it was measured that it exceeded 28 °C, the landing was surely executed. By realizing this system, it is possible to deal with the troubles of the motor, which is supposed to occur when the drones are able to fly for a long time in the future. We also consider to use deep learning methods [6, 7] to select the threshold temperature.

Acknowledgments This work was supported by JSPS KAKENHI (15F15077), Leading Initiative for Excellent Young Researcher (LEADER) of Ministry of Education, Culture, Sports, Science and Technology-Japan (16809746), Research Fund of Chinese Academy of Sciences (No.MGE2015-KG02), Research Fund of State Key Laboratory of Marine Geology in Tongji University (MGK1608), Research Fund of State Key Laboratory of Ocean Engineering in Shanghai Jiaotong University (1315; 1510), and Research Fund of The Telecommunications Advancement Foundation.

References

1. NTT DATA, Started providing IOT solution to detect abnormality from working sound, NTT DATA,2016/10/3. https://latte.la/column/43321221
2. Nippon Avionics Co., (1999) Ltd, K-1. Motor abnormal temperature detection system |Infrared thermography |Nippon Avionics Nippon Avionics Co., Ltd. http://www.avio.co.jp/products/infrared/solution/k-heavy/k-01.html
3. Oriental Motor Co., Ltd, Design life of motor |Technical data|. Oriental Motor Co., Ltd. https://www.orientalmotor.co.jp/tech/reference/life01/
4. Hirozawa Seiki Seisakusho Co., Ltd, Metal Feather Fan Sensor, Hirozawa Seiki Seisakusho Co., Ltd. http://www.hirosawaseiki.co.jp/ikurafan/pages/Cat/IkuraFan1406_084.pdf
5. Yuuto, T.: Detection and deterioration diagnosis of electrical equipment based on current waveform pattern analysis, 2013/11/29. http://vision.kyoto-u.ac.jp/japanese/happyou/pdf/Y-Tamura_ASN_20131129.pdf
6. Li, Y., Lu, H., Li, J., Li, X., Li, Y., Serikawa, S.: Underwater image de-scattering and classification by deep neural network. Comput. Electr. Eng. **54**, 68–77 (2016)
7. Lu, H., Li, B., Zhu, J., Li, Y., Li, Y., Xu, X., He, L., Li, X., Li, J., Serikawa, S.: Wound intensity correction and segmentation with convolutional neural networks. Concurr. Comput. Pract. Exper. **29**(6), e3927 (2017)

Underwater Light Field Depth Map Restoration Using Deep Convolutional Neural Fields

Huimin Lu, Yujie Li, Hyoungseop Kim and Seiichi Serikawa

Abstract Underwater optical images are usually influenced by low lighting, high turbidity scattering and wavelength absorption. In order to solve these issues, a great deal of work has been used to improve the quality of underwater images. Most of them used the high-intensity LED for lighting to obtain the high contrast images. However, in high turbidity water, high-intensity LED causes strong scattering and absorption. In this paper, we firstly propose a light field imaging approach for solving underwater depth map estimation problems in low-intensity lighting environment. As a solution, we tackle the problem of de-scattering from light field images by using deep convolutional neural fields in depth estimation. Experimental results show the effectiveness of the proposed method through challenging real world underwater imaging.

Keywords Light field camera · Underwater imaging · Deep convolutional neural fields

1 Introduction

Light field or plenoptic camera has been applied in many computer vision applications since it appeared in 1966 [1]. Light field camera, such as Lytro [2], Raytrix [3] and Profusion [4], measures both geometric and color information. It can be applied in extreme environment instead of other RGB-D cameras, e.g. in bright/low light or underwater [5]. Although Ref. [5] firstly introduced to use the light field camera for underwater image denoising, this article is the first time focusing on solving the underwater image de-scattering and de-absorption. Traditionally, there

H. Lu (✉) · H. Kim · S. Serikawa
Kyushu Institute of Technology, Fukuoka, Japan
e-mail: luhuimin@ieee.org

Y. Li
Yangzhou University, Jiangsu, China
e-mail: liyujie@yzu.edu.cn

© Springer International Publishing AG 2018
H. Lu and X. Xu (eds.), *Artificial Intelligence and Robotics*,
Studies in Computational Intelligence 752,
https://doi.org/10.1007/978-3-319-69877-9_33

305

are three basic steps when we apply the light field cameras, that is, light field data acquisition, light field calibration, and light field applications.

For light field acquisition, Wilburn et al. [6] designed a bulky camera array. This light field imaging system can capture a light field image with high spatial and angular resolution. Veeraraghavan et al. [7] proposed a light field acquisition method, which was based on a transparent make adjacent in the front of the camera. Liang et al. [8] used programmable aperture patterns to encode angular information of light rays. Taguchi et al. [9] designed a hemispherical mirrors lens for widening the view of light field camera. However, these light field imaging systems are usually huge and expensive.

In 2005, Ng et al. [10] presented a hand-held light field camera using a micro-lens array. The attached micro-lens estimates the directional distribution of incoming rays. Georgiev et al. [11, 12] modified the Ng's model, and improved the system to capture a higher spatial resolution light field image.

The rest of the paper is organized as the following. In Sect. 2, we introduce the depth estimation method for light field cameras.

2 Depth Image Estimation Methods

Many studies have been researched to estimate the depth information from the light field image. Wanner et al. [13] proposed to use a vibrational global optimization scheme to label the depth level globally consistent. Tao et al. [14] presented an algorithm that computes the depth by combining both defocus and correspondence depth cues. They also proposed an iterative approach to estimate the depth information as well as remove the specular component [15]. Jeon et al. [16] introduced an algorithm that estimates the multi-view correspondences with sub-pixel accuracy with the cost volume. Wang et al. [17] proposed a depth estimation method that can detect the occlusion area. Williem et al. [18] developed an algorithm for estimating the depth from occlusion and noisy environment. Kalantari et al. [19] proposed two sequential convolutional neural networks to model the components and synthesize new views from a sparse set of input views. Wang et al. [20] proposed to use the low resolution depth map to generate a depth map of the main scene. Wang et al. [21] derived a spatially-varying BRDF-invariant theory for recovering 3D shape and reflectance from light field cameras.

ToF camera based depth map refinement methods can be divided into two categories: Markov Random Field (MRF)-based methods and Advanced Filtering (AF) methods. Diebel et al. [22] used a two-layer MRF to model the correlation of range measurements and refine the depth map with the conjugate gradient method. Huhle et al. [23] proposed a third-layer MRF to improve the depth map. Garro et al. [24] proposed a graph-based segmentation method to interpolate the missing depth information. Yang et al. [25] proposed a hierarchical joint bilateral filtering method for depth map upsampling. Zhu et al. [26] designed a dynamic MRFs for improving the accuracy of depth map both in the spatial and temporal domain. He et al. [27]

proposed a guided image filter to smooth the depth information with reference image. Lu et al. [28] incorporated amplitude values of ToF camera and designed a data term to fit the characteristics of depth map. Park et al. [29] considered to use a non-local term and weighted color image filtering to recover the depth information. Aodha et al. [30] used a generic database of local patches to increase the resolution of depth map. Min et al. [31] proposed a weighted mode filtering to modify the histogram of depth map. Lu et al. [32] proposed a local multipoint regression method to estimate the depth information of each point. Ferstl et al. [33] used an anisotropic diffusion tensor obtained from a reference image. Liu et al. [34] investigated a geodesic distance to compute the filtering coefficient of similarity pixels.

3 Proposed Method

In this paper, we combine the underwater dehazing method [35, 36] and deep convolutional neural fields depth estimation method [37] to estimate the exactly depth map. According to [36], the underwater imaging model can be defined as

$$I^c(x) = J^c(x)T^c(x) + (1 - T^c(x))A^c, c \in \{r, g, b\}. \tag{1}$$

where we suppose the light $J^c(x)$ reflected from point x, the inhomogeneous background A^c represents the ambient light and $T^c(x)$ is the transmission map. Then, through locally adaptive cross filter-based descattering, we can remove the high turbid artifacts in underwater image.

After descattering, we use the deep convolutional neural fields to estimate the depth map. Because the underwater light field depth dataset is hardly to build, we use the NYU v2 for training. This method uses a superpixels segmentation and a fusion method of CNN and continuous CRF, which is denoted as DCNF. The descattered image is over-segmented into superpixels. Then, use a window to crop the image to many patches, which is denoted as superpixel image patch. All of the superpixel image patches are resized and feed to a CNN which is composed of 5 convolutional layers and 4 fully connected layers. At the meantime, a pair of neighbor superpixels are feed to a fully connected layer. The outputs of depth map is calculated the minimization of negative log-likelihood.

4 Experimental Results

In this paper, we used the water tank with the size of 115 cm × 35 cm × 45 cm in darkroom. The Lytro Illum camera was set in front of the water tank with the distance of 75 cm. In order to simulate the deep-sea environment, we put the black

Fig. 1 Imaging equipment
setting

plastic plate around the water tank. Figure 1 shows the image equipment setting of this experiment.

We put three toys in water tank with different distance from the camera. In the experiment, we added the coffee and milk mixed liquid into the water. The clean water becoming turbid when gradually increase the dose.

In this experiment, we compared the proposed method with [16], which is the state of the art method in the world. From Fig. 2b to d, we can find that there are some mis-estimations. Compare with the Jeon's method [16], the proposed method performs well. However, there also have some drawbacks, such as errors of depth information of the last toy. In future, we will consider to use the other deep learning methods, brain intelligence [38, 39] to improve the accuracy of depth estimation.

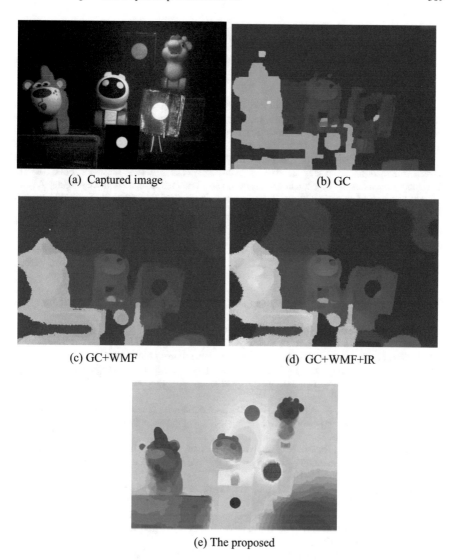

(a) Captured image   (b) GC

(c) GC+WMF   (d) GC+WMF+IR

(e) The proposed

Fig. 2 Estimated depth image of different algorithms. **a** Captured RGB image; **b** Result of multi-label optimization with Graph-cuts; **c** Result of Graph-cuts and weighted median filter; **d** Result of Graph-cuts, weighted median filter and iterative refinement; **e** The proposed method

5 Conclusion

This paper proposed an underwater image depth estimation method, which is based on light field camera and deep convolutional neural fields. This is the first work for solving underwater light field image depth map estimation problems with deep learning technologies. Compared with the traditional methods, the proposed method achieve better depth information. However, there also exist some errors of the estimation result. In future, we will build an underwater light field dataset, we also will to use the other deep learning methods instead of DCNF to improve the performance of this system.

Acknowledgements This work was supported by Leading Initiative for Excellent Young Researcher (LEADER) of Ministry of Education, Culture, Sports, Science and Technology-Japan (16809746), Grants-in-Aid for Scientific Research of JSPS (17K14694), Research Fund of Chinese Academy of Sciences (No.MGE2015KG02), Research Fund of State Key Laboratory of Marine Geology in Tongji University (MGK1608), Research Fund of State Key Laboratory of Ocean Engineering in Shanghai Jiaotong University (1315;1510), Research Fund of The Telecommunications Advancement Foundation, and Fundamental Research Developing Association for Shipbuilding and Offshore. We also thank Dr. Donald Dansereau at Stanford University for contributing the imaging equipment setting.

References

1. Ligten, R.: Influence of photographic film on wavefront reconstruction. J. Opt. Soc. Am. **56**, 1009–1014 (1966)
2. The Lytro Camera. http://www.lytro.com/
3. Raytrix: 3D light field camera technology. http://www.raytrix.de/
4. ProFusion. http://www.viewplus.co.jp/product/camera/profusion25.html/
5. Dansereau, D., Bongiorno, D., Pizarro, O., Williams, S.: Light field image denoising using a linear 4D frequency-hyperfan all-in-focus filter. In: Proceedings SPIE Computational Imaging XI, Feb 2013
6. Wilburn, B., Joshi, N., Vaish, V., Talvala, E., Antunez, E., Barth, A., Adams, A., Horowitz, M., Levoy, M.: High performance imaging using large camera arrays. ACM Trans. Graphics **24**(3), 765–776 (2005)
7. Veeraraghavan, A., Raskar, R., Agrawal, A., Mohan, A., Tumblin, J.: Dappled photography: mask enhanced cameras for heterodyned light fields and coded aperture refocusing. ACM Trans. Graphics **26**(3), 1–10 (2007)
8. Liang, C., Lin, T., Wong, B., Liu, C., Chen, H.: Programmable aperture photography: multiplexed light field acquisition. ACM Trans. Graphics **27**(3), 1–10 (2008)
9. Taniguchi, Y., Agrawal, A., Veeraraghavan, A., Ramalingam, S., Raskar, R.: Axial-cones: modeling spherical catadioptric cameras for wide-angle light field rendering. ACM Trans. Graphics **29**(6), 1–10 (2010)
10. Ng, R., Levoy, M., Bredif, M., Duval, G., Horowitz, M., Hanrahan, P.: Light field photography with a hand-held plenoptic camera. Stanford University Computer Science and Technical Report, vol. 2, no. 11, 2005
11. Georgiev, T., Lumsdaine, A.: Reducing plenoptic camera artifacts. In: Computer Graphics Forum, vol. 29, no. 6, pp. 1955–1968 (2010)
12. Lu, H., Li, B., Zhu, J., Li, Y., Li, Y., Xu, X., He, L., Li, X., Li, J., Serikawa, S.: Wound intensity correction and segmentation with convolutional neural networks. Concurr. Comput. Pract. Exp. **27**(9), 1–10 (2017)
13. Wanner, S., Goldluecke, B.: Globally consistent depth labeling of 4D light fields. In Proceedings of CVPR2012, pp. 41–48 (2012)
14. Tao, M., Hadap, S., Malik, J., Ramamoorthi, R.: Depth from combining defocus and correspondence using light-field cameras. In Proceedings of IEEE ICCV2013, pp. 673–680 (2013)

15. Tao, M., Wang, T., Malik, J., Ramamoorthi, R.: Depth estimation for glossy surfaces with light-field cameras. In: Workshop on Light Fields for Computer Vision, ECCV (2014)

16. Jeon, H., Park, J., Choe, G., Park, J., Bok, Y., Tai, Y., Kweon, I.: Accurate depth map estimation from a lenslet light field camera. In: Proceedings of CVPR2015, pp. 1547–1555 (2015)

17. Wang, W., Efros, A., Ramamoorthi, R.: Occlusion-aware depth estimation using light-field cameras. In Proceedings of ICCV2015, pp. 3487–3495 (2015)

18. Williem, W., Park, I.: Robust light field depth estimation for noisy scene with occlusion. In: Proceedings Of CVPR2016, pp. 4396–4404 (2010)

19. Kalantari, N., Wang, T., Ramamoorthi, R.: Learning-based view synthesis for light field cameras. In Proceedings of SIGGRAPH Asia (2016)

20. Wang, T., Srikanth, M., Ramamoorthi, R.: Depth from semi-calibrated stereo and defocus. In: Proceedings of CVPR (2016)

21. Wang, T., Chandraker, M., Efros, A., Ramamoorthi, R.: SVBRDF-invariant shape and reflectance estimation from light-field cameras. In Proceedings of CVPR (2016)

22. Diebel, J., Thrun, S.: An application of Markov radom fields to range sensing. Adv. Neural. Inf. Process. Syst. **18**, 291 (2005)

23. Huhle, B., Fleck, S., Schilling, A.: Integrating 3D time-of-flight camera data and high resolution images for 3DTV applications. In: Proceedings of 3DTV, pp. 1–4 (2007)

24. Garro, V., Zanuttigh, P., Cortelazzo, G.: A new super resolution technique for range data. In Proceedings of Associazione Gruppo Telecomunicazionie Tecnologie dell Informazione (2009)

25. Yang, Q., Tan, K., Culbertson, B., Apostolopoulos, J.: Fusion of active and passive sensors for fast 3D capture. In Proceedings of IEEE International Workshop on Multimedia Signal Processing, pp. 69–74 (2010)

26. Zhu, J., Wang, L., Gao, J., Yang, R.: Spatial-temporal fusion for high accuracy depth maps using dynamic MRFs. IEEE Trans Pattern Anal. Mach. Intell. **32**(5), 899–909 (2010)

27. He, K., Sun, J., Tang, X.: Guided image filtering. In: Proceedings of ECCV, pp. 1–14 (2010)

28. Lu, J., Min, D., Pahwa, R., Do, M.: A revisit to MRF-based depth map super-resolution and enhancement. In: Proceedings of IEEE ICASSP, pp. 985–988 (2011)

29. Park, J., Kim, H., Tai, Y., Brown, M., Kweon, I.: High quality depth map upsampling for 3D-TOF cameras. In: Proceedings of ICCV, pp. 1623–1630 (2011)

30. Aodha, O., Campbell, N., Nair, A., Brostow, G.: Patch based synthesis for single depth image super-resolution. In: Proceedings of ECCV, pp. 71–84 (2012)

31. Min, D., Lu, J., Do, M.: Depth video enhancement based on joint global mode filtering. IEEE Trans. Image Process. **21**(3), 1176–1190 (2012)

32. Lu, J., Shi, K., Min, D., Lin, L., Do, M.: Cross-based local multipoint filtering. In Proceedings of CVPR, pp. 430–437 (2012)

33. Ferstl, D., Reinbacher, C., Ranftl, R., Ruther, M., Bischof, H.: Image guided depth upsampling using anisotropic total generalized variation. In: Proceedings of ICCV, pp. 993–1000 (2013)

34. Liu, M., Tuzel, O., Taguchi, Y.: Joint geodesic upsampling of depth images. In Proceedings of CVPR, pp. 169–176 (2013)

35. Serikawa, S., Lu, H.: Underwater image dehazing using joint trilateral filter. Comput. Electr. Eng. **40**(1), 41–50 (2014)

36. Lu, H., Li, Y., Zhang, L., Serikawa, S.: Contrast enhancement for images in turbid water. J. Opt. Soc. Am. **32**(5), 886–893 (2015)

37. Liu, F., Shen, C., Lin, G., Reid, I.: Learning depth from single monocular images using deep convolutional neural fields. IEEE Trans. Pattern Anal. Mach. Intell. **38**(10), 2024–2038 (2016)

38. Lu, H., Li, Y., Chen, M., Kim, H., Serikawa, S.: Brain Intelligence: Go Beyond Artificial Intelligence. arXiv:1706.01040 (2017)
39. Lu, H., Zhang, Y., Li, Y., Zhou, Q., Tadoh, R., Uemura, T., Kim, H., Serikawa, S.: Depth map reconstruction for underwater Kinect camera using inpainting and local image mode filtering. IEEE Access **5**(1), 7115–7122 (2017)

Image Processing Based on the Optimal Threshold for Signature Verification

Mei Wang, Min Sun, Huan Li and Huimin Lu

Abstract To realize the rapid and accurate signature verification, a new image processing method is developed on the basis of the optimal threshold algorithm. Firstly, the improved Gaussian filtering (IGF) algorithm is developed for the signature image to remove the noises. Secondly, the optimal threshold (OT) algorithm is developed to find the optimal threshold for the signature image segmentation. Finally, the deep learning method of a convolutional neural network is used to verify the signature image. It is experimentally proved that the IGF algorithm can get a better filtering effect, and the OT algorithm can obtain the better segmentation result, and the system has the better recognition accuracy.

Keywords Signature verification · Image segmentation · Convolutional neural network · Threshold · Filtering

1 Introduction

The rapid development of modern information technologies have led people to be gradually used to the life style of the various digital information network. In this background, identity authentication has become an indispensable part of people's daily life. The signature is the human's characteristic which is currently used with a

M. Wang · M. Sun · H. Li
College of Electrical and Control Engineering, Xi'an University of Science and Technology, Xi'an, China
e-mail: wangm@xust.edu.cn

M. Sun
e-mail: 15129360862@163.com

H. Li
e-mail: leeo12306@163.com

H. Lu (✉)
Kyushu Institute of Technology, 1-1 Sensui, Tobata, Kitakyushu 804-8550, Japan
e-mail: luhuimin@ieee.org

© Springer International Publishing AG 2018
H. Lu and X. Xu (eds.), *Artificial Intelligence and Robotics*,
Studies in Computational Intelligence 752,
https://doi.org/10.1007/978-3-319-69877-9_34

313

higher utilization ratio in many areas, especially in the mobile terminal system. The research on signature authentication has become a hot topic for the mobile terminal systems.

For the signature authentication, researches proposed the training strategy of the deep convolution neural networks with distorted samples to reduce the distortion and to increase the generalization of the neural network [1, 2]. To get the best match, researchers developed the off-line signature alignment verification method on the basis of Gaussian mixture model [3]. Besides, scientists proposed the offline signature verification method on the basis of the fractional fusion of a distance with the centroid orientation [4]. Scientists also presented a classification method of the multi-task metric learning. This method used the true and the fake samples to obtain the similarity and dissimilarity knowledge [5]. In addition, scientists encoded the fine geometric structure by using the mesh template, and split the area in the subset for offline signature analysis and verification [6]. Furthermore, scientists studied DTW method for the in-air signature verification based on the video [7].

On the other hand, researchers proposed an online signature verification technique based on discrete cosine transform and sparse representation [8]. Also, they proposed the group sparse representation for a separate feature-level fusion mechanism and an integrates multiple feature representation [9]. In addition, the probabilistic class structure [10] and the robust facial expressions [11] as well as the sparse exponents batch processing method [12] were developed.

Different from the the above methods, this paper focus on the image segmentation based on the optimal threshold for signature verification. The organization of this paper is as follows: Sect. 2 develops the improved Gaussian filtering (IGF) algorithm. Section 3 develops the optimal threshold (OT) segmentation algorithm. Section 4 describes the convolutional neural network to verify the signature. Finally, Sect. 5 is the conclusions.

2   The Improved Gaussian Filtering (IGF) of Signature Image

In order to remove the noises and make the signature image smooth, the processing filtering is needed firstly for the signature authentication [13–15]. Filtering is the convolution process of the input signature image with a convolution core. The commonly used filtering method is the Gaussian filtering method.

In this paper, the central point of the Gaussian filtering method is changed to a suitable point to develop the IGF algorithm. Figure 1 shows the effect comparisons of the IGF algorithm with the traditional methods.

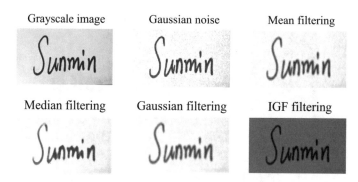

Fig. 1 The effect comparisons of the IGF algorithm with the traditional methods

3   The Developed Optimal Threshold (OT) Algorithm

Image segmentation is to set all the pixel value to be 0 or 1. Meanwhile, the pixel positions keep invariant. This operation is for simplifying the post processing. The segmentation threshold influences the result dramatically. Figure 2 shows the gray signature image and the histogram program as well as the segmentation with different thresholds.

To obtain a better segmentation effect, we develops the OT algorithm to segment the signature image. For a image, the gray degree histogram have 2 peaks. One is for the background, and the other is for the object. There must be a minimum point P between the 2 peaks. At point P, the gray degree value corresponds to the minimum histogram value Q. We select this the gray degree value Q to be the optimal threshold for the segmentation of the filtered signature image (Fig. 3).

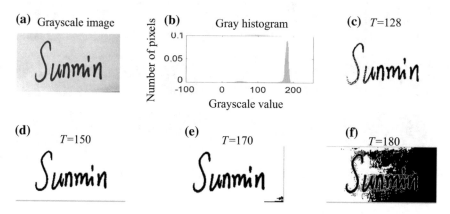

Fig. 2 Gray signature image and the histogram program as well as the segmentation with different thresholds

Grayscale image   Entropy method

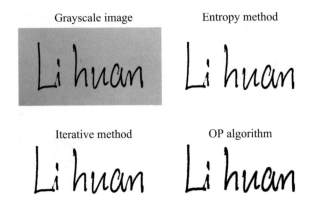

Iterative method   OP algorithm

Fig. 3 The effect comparisons of the OT algorithm with the traditional methods

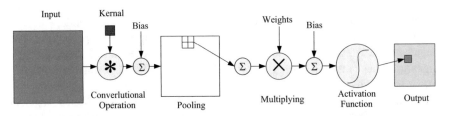

Fig. 4 The process of converlution and sampling of the convolutional neural network

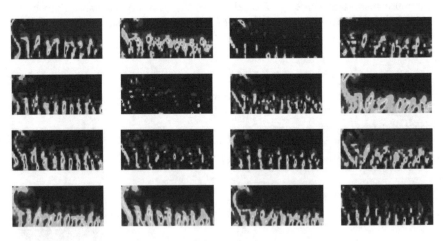

Fig. 5 The signature feature image of the convolutional neural network

4 Signature Image Verification Using a Convolutional Neural Network

In this paper, the convolutional neural network is used to extract the signature and verify the signature. Figure 4 is the process of converlution and sampling of the convolutional neural network. Figure 5 is the signature feature image of the convolutional neural network. Figure 6 shows the accuracy and speed as well as iterative error of CNN using different activation functions. It can be seen that the activation function Relu has the better performance for the signature verification. Therefore, it is selected as the activation function of the convolutional neural network. Figure 7 is the accuracy comparison and the time comparison. It can be seen that the convolutional neural network (CNN) has the relative higher accuracy and the relative lower time consuming.

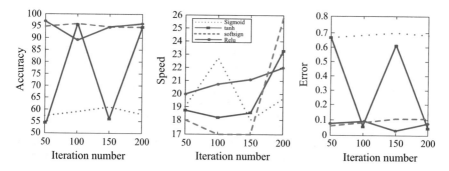

Fig. 6 The accuracy and speed as well as iteration error of CNN using different activation functions

Fig. 7 The accuracy comparison and the time comparison

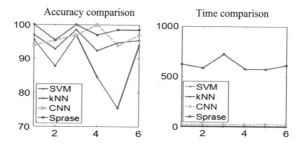

5 Conclusions

This paper highlights the a new image processing method to realize the filtering and segmentation of the signature image. In this image, the IGF algorithm is developed for the signature image to remove the noises. Then, the OT algorithm is developed to find the optimal threshold for the signature image segmentation. In addition, the convolutional neural network is trained to verify the signature image. It is experimentally proved that the IGF algorithm can get a better filtering effect, and the OT algorithm can obtain the better segmentation result, and the signature recognition system has a better recognition accuracy.

Acknowledgements This research was sponsored by the Natural Science Foundation of China (51405381), Key Scientific and Technological Project of Shaanxi Province (2016GY-040), and the Science Foundation of Xi'an University of Science and Technology (104-6319900001).

References

1. Lai, S., Jin, L., Yang, W.: Toward high-performance online HCCR: a CNN approach with drop distortion, path signature and spatial stochastic max-pooling. Pattern Recogn. Lett. **89**, 60–66 (2017)
2. Wang, M., Guo, L., Chen, W.Y.: Blink detection using Adaboost and contour circle for fatigue recognition. Comput. Electr. Eng. **58**, 502–512 (2016)
3. Xia, X., Chen, Z., Luan, F., Song, X.: Signature alignment based on GMM for on-line signature verification. Pattern Recogn. **65**, 188–196 (2017)
4. Manjunatha, K.S., Manjunath, S., Guru, D.S., Somashekara, M.T.: Online signature verification based on writer dependent features and classifiers. Pattern Recogn. Lett. **80**, 129–136 (2016)
5. Soleimani, A., Araabi, B.N., Fouladi, K.: Deep multitask metric learning for offline signature verification. Pattern Recogn. Lett. **80**, 84–90 (2016)
6. Zois, E.N., Alewijnse, L., Economou, G.: Offline signature verification and quality characterization using poset-oriented grid features. Pattern Recogn. Lett. **54**, 162–177 (2016)
7. Sharma, A., Sundaram, S.: An enhanced contextual DTW based system for online signature verification using vector quantization. Pattern Recogn. Lett. **84**, 22–28 (2016)
8. Fang, Y., Kang, W., Wu, Q., Tang, L.: A novel video-based system for in-air signature verification. Comput. Electr. Eng. **57**, 1–14 (2017)
9. Liu, Y., Yang, Z., Yang, L.: Online signature verification based on DCT and sparse representation. IEEE Trans. Cybern. **45**, 2498 (2015)
10. Goswami, G., Mittal, P., Majumdar, A., Vatsa, M., Singh, R.: Group sparse representation based classification for multi-feature multimodal biometrics. Inf. Fusion **32**, 3–12 (2016)
11. Shao, Y., Sang, N., Gao, C., Ma, L.: Probabilistic class structure regularized sparse representation graph for semi-supervised hyperspectral image classification. Pattern Recogn. **63**, 102–114 (2017)
12. Ouyang, Y., Sang, N., Huang, R.: Accurate and robust facial expressions recognition by fusing multiple sparse representation based classifiers. Elsevier Sci. Publ. B **149**, 71–78 (2015)

13. Lu, H., Li, Y., Chen, M., Kim, H., Serikawa, S.: Brain intelligence: go beyond artificial intelligence. Mob. Netw. Appl. 1–10 (2017)
14. Lu, H., Li, Y., Mu, S., Wang, D., Kim, H., Serikawa, S.: Motor anomaly detection for unmanned aerial vehicles using reinforcement learning. IEEE Internet of Things J. https://doi.org/10.1109/JIOT.2017.2737479
15. Serikawa, S., Lu, H.: Underwater image dehazing using joint trilateral filter. Comput. Electr. Eng. **40**(1), 41–50 (2014)

Fault Location Without Wave Velocity Influence Using Wavelet and Clark Transform

Mei Wang, Changfeng Xu and Huimin Lu

Abstract To eliminating the influence of traveling wave velocity on the fault location accuracy of the power line, a fault location method without wave velocity influence using Wavelet and Clark transform is developed in this paper. On the basis of the reflection characteristic analysis of fault traveling wave, the aerial mode component of the voltage traveling wave is obtained by Clark transform. Then, the fault time when the first three mode maximum appear on the measuring end is determined by the combination of the aerial mode component with Wavelet transform. Furthermore, a set of equations that is composed by the fault time and the fault distance is formed. This method is independent of the wave velocity and is not affected by the traveling wave velocity theoretically. The simulation results show that the developed method can reduce the relative error of the fault distance by less than 5%, which proves the validity and the accuracy of this method.

Keywords Traveling wave velocity · Fault location · Clark transform Wavelet transform · Mode maximum

1 Introduction

Power line play the extremely important role in our society. The research on the fault location method of the power line attract many scientist to study. The existing fault location algorithm using the double ended traveling wave is a commonly used

M. Wang · C. Xu
College of Electric and Control Engineering, Xi'an University of Science and Technology, Xi'an, China
e-mail: wangm@xust.edu.cn

C. Xu
e-mail: 1281796036@qq.cn

H. Lu (✉)
Kyushu Institute of Technology, 1-1 Sensui, Tobata, Kitakyushu 804-8550, Japan
e-mail: luhuimin@ieee.org

© Springer International Publishing AG 2018
H. Lu and X. Xu (eds.), *Artificial Intelligence and Robotics*,
Studies in Computational Intelligence 752,
https://doi.org/10.1007/978-3-319-69877-9_35

method for the fault location. This method is less affected by the power line parameters, the system operation mode, the fault type and the fault impedance. It used the time when the fault traveling wave arrives at the measuring point. In literature [1], the multiple scale analysis and the mutative and singular information representation of the wavelet transform is used to find the start point and the inflection point, then the fault distance is calculated. In literature [2], the fault location algorithm uses the global positioning system to determine the power cable fault distance on the base of the discrete wavelet transform. However, the system cost is huge and time consuming is longer. In literature [3], a double ended traveling wave algorithm independent the wave velocity is proposed, but the time of the fault reflection to the opposite bus is easily confused with the time that the fault is reflected to the initial bus. In addition, it is proved by the research [4] that the uncertainty of the wave velocity will affect the accuracy of traveling wave fault location.

In this paper, we use wavelet transform to detect the size of singularity and the time of singular point [5]. We solve a set of equations consist of distance, time and wave velocity, and get a double end traveling wave fault location method that independent wave velocity. Eventually, the power line fault distance is calculated accurately.

2 Construction of the Aerial Mode with a Stable Wave Velocity

When the traveling wave propagates in the three phase lines where the neutral point is not grounded, there is no single phase traveling wave due to the coupling among the three lines [6]. Applying traveling wave method to the three phase system, and the phase voltage and the phase current must be decoupled by the phase mode transform. Therefore, the measured signal must be transformed into a mode signal. Using the Clark transform to convert the phase a, b, c into 0, α, β mode, there is no the coupling among the modes, and the modes are independent to each other. In the current relay protection [7], the application of the phase mode transform technology is becoming more and more popular. The zero mode is the earth vector, and its circuit parameters are influenced by the ground conditions and soil resistivity and the other factors. The speed is extremely unstable, so it's unsuitable for the fault location. Most of the fault analysis using α and β mode components as below.

$$T = \begin{bmatrix} \alpha \\ \beta \\ 0 \end{bmatrix} = \frac{2}{3} \begin{bmatrix} 1 & -\frac{1}{2} & -\frac{1}{2} \\ 0 & \frac{\sqrt{3}}{2} & -\frac{\sqrt{3}}{2} \\ \frac{1}{2} & \frac{1}{2} & \frac{1}{2} \end{bmatrix} \begin{bmatrix} a \\ b \\ c \end{bmatrix} \tag{1}$$

where a, b, c represents the three phase voltage values respectively, and the α, β, 0 are the modes.

3   Fault Information Extraction by Wavelet Transform

Wavelet transform has a good partial characteristics in time-frequency domain. In order to locate the fault point accurately, this method solves the problems of the off line fault monitoring and the big error on the base of wavelet transform. It is important to analyze signal singularity, and this is effective to find the starting point of the reflection pulse [8]. At the same time, Wavelet transform can effectively eliminate noise. These attributes are very suitable for analyzing transient traveling waves [9].

In the method of the fault location using traveling wave, the wavelet transform will extract the fault information from the acquisition signal [10]. Since the fault is usually manifested as a signal mutation. It is necessary to extract the instantaneous and abrupt components of the non-stationary signal [11]. Therefore, the selection of wavelet basis will greatly affect the analysis of the fault signal. Compared with other wavelet series, the db wavelet series can extract fault information precisely, so the db wavelet is very suitable to identify instantaneous fault signals [12–15].

4   Derivation of the Fault Location Formula

Assume the total length of the line MN is L, and the distance between M and fault point is D. The time of initial traveling wave reach to the bus M is t_{M1}, and its wave velocity is V_{M1}. The time of second traveling wave reach to M is t_{M2}, and its wave velocity is V_{M2}. The time of the third traveling wave reach the bus M terminal is t_{M3}, and its wave velocity is V_{N1}.

When a fault occurs in the first half of the line, the equations of traveling wave is as follow.

$$(t_{M1} - t_0) \times V_{M1} = D \tag{2}$$

$$(t_{M2} - t_{M1} - t_0) \times V_{M2} = 2D \tag{3}$$

$$(t_{M3} - t_0) \times V_{N1} = 2L - D \tag{4}$$

where V_{M1} is aV_0, V_{M2} is bV_0, and V_{N1} is cV_0. Meanwhile, $0 < a, b, c < 1$, and V_0 is the ultimate wave velocity.

Solve these equations, we can get the formula of double ended traveling wave in first half.

$$D = \frac{ab(t_{M2} - 2t_{M1})}{At_{M1} + Bt_{M2} + Ct_{M3}} \tag{5}$$

where $A = -2ab - (2a + b)c$, $B = ab + bc$, $C = (2a - b)c$.

If a fault occurs in the latter half of the line, the equations of traveling wave is as follow.

$$(t_{M1} - t_0) \times V_{M1} = D \tag{6}$$

$$(t_{M2} - t_0) \times V_{N2} = 2L - D \tag{7}$$

$$(t_{M3} - t_{M1} - t_0) \times V_{M2} = 2D \tag{8}$$

where V_{M1} is aV_0, and V_{M2} is bV_0, and V_{N1} is cV_0. Meanwhile, $0 < a, b, c < 1$. V_0 is the ultimate wave velocity.

Solve these equations, we can get the formula of double ended traveling wave in latter half.

$$D = \frac{ab(t_{M3} - 2t_{M1})}{A_1 t_{M1} + B_1 t_{M2} + C_1 t_{M3}} \tag{9}$$

where $A_1 = -2ab - (2a + b)c$, $B_1 = (2a - b)c$, $C_1 = ab + bc$.

5 Experiment Environment and Discussion

For a simulation experiment, a power line model is established by the software MATLAB. The length of line is 20 km, and the voltage is 6 kV. The specific parameters of line are as the follows. The positive sequence resistance $R_1 = 0.01237$ Ω/km. The zero sequence resistance $R_0 = 0.3864$ Ω/km. The positive sequence and the zero sequence inductance $L_1 = 0.9337$ mH/km and $L_0 = 4.1264$ mH/km, The positive sequence and the zero sequence capacitance $C_1 = 12.74$ uF/km and $C_0 = 7.751$ uF/km. The starting fault time is 0.03 s. The duration time of the fault is 0.01 s.

If the single phase ground fault occurs in the line, the fault distance is 5 km. The three phase fault voltage transform by Clark mode transform will get U_0, U_α, U_β, U_α waveform as shown in Fig. 1. The transform U_α mode by wavelet transform will obtain Fig. 2a. After the denoising by the mode maximum and the appropriate threshold processing, the mode maximum of each layer is obtained, as shown in Fig. 2b. We can clearly see three moments of the wave heads to the front end of the starting bus M.

If a fault occurs in 13 km, wavelet transform is utilized to obtain the mode maximum and get the fault times, $t_{M1} = 0.00992$ s, $t_{M2} = 0.01001$ s, $t_{M3} = 0.01011$ s. These fault times are substituted to Eq. (9), and the fault distance is calculated, $D = 13.07$ km, and the relative error is $e = 0.59\%$. If we adopt traditional method, the fault distance is $D = 13.27$ km.

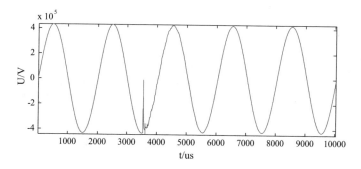

Fig. 1 Waveform of U_α

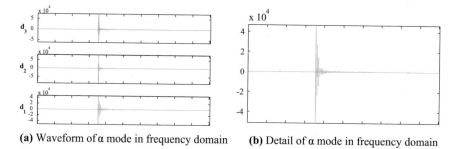

(a) Waveform of α mode in frequency domain   **(b)** Detail of α mode in frequency domain

Fig. 2 Waveform in frequency domain

6 Conclusion

A method that avoid the effect of wave velocity and combine wavelet analysis with Clark transform is developed to position the cable fault. The Clark transform is used to uncouple the phase signal into the independent mode, and then the fault time is obtained by the mode maximum of wavelet transform. A set of equations consist of the distance, the time and the velocity is obtained, and the wave velocity is calculated by these equations. The simulation result shows that the relative error of cable fault is less than 5%. The developed algorithm unaffected by the wave velocity, which has a widely practical value.

Acknowledgements This research was sponsored by the Natural Science Foundation of China (51405381), Key Scientific and Technological Project of Shaanxi Province (2016GY-040), and the Science Foundation of Xi'an University of Science and Technology (104-6319900001).

References

1. Min, X., Yan-wei, Y., Ke-yun, S.: Cable short open-circuit fault location based on wavelet transform. Power Syst. Protect. Control **4**, 41–45 (2013)
2. Mosavi, M.-R., Tabatabaei, A.: Traveling-wave fault location techniques in power system based on wavelet analysis and neural network using GPS timing. Wireless Pers. Commun. **86**, 835–850 (2016)
3. Xiaoguang, Y., Linlin, S., Zhi, Y.: Study of fault locating for transmission line double terminal traveling waves unrelated to wave speed **39**, 35–39 (2011)
4. Wen, S., Yadong, L., Ge-hao, S., Xuri, S.: Influence of operating state of overhead transmission line on traveling wave velocity. In: Proceedings of the CSU-EPSA, vol. 26, pp. 12–16 (2014)
5. XinZhou, D., ShenXing, S., Tao, C.: Optimizing solution of fault location using single terminal quantities. Sci. China Ser. E Technol. Sci. **51**, 761–772 (2008)
6. Lin, X., Zhao, F., Gang, W.: Universal wavefront positioning correction method on traveling wave based fault location algorithm. Power Deliv. **27**, 1601–1610 (2012)
7. Huifeng, Z., Xiangjun, Z.: A new network-based algorithm for transmission line fault location with traveling wave. Autom. Electric Power Syst. **19**, 93–99 (2013)
8. Mosavi, M.R., Tabatabaei, A.: Wavelet and neural network-based fault location in power systems using statistical analysis of traveling wave. Arab. J. Sci. Eng. **8**, 6207–6214 (2014)
9. Chen, L., Yu, Y.: An improved fault detection algorithm based on wavelet analysis and kernel principal component analysis. IEEE Control Decis. Conf. (CCDC) **26**, 1723–1726 (2010)
10. Shenxing, S., Xinzhou, D., Shuangxi, Z.: New principle to identify the second reverse traveling wave generated by single-phase-to-ground fault. Autom. Electric Power Syst. **30**, 41–44 (2006)
11. Wang, X., Wang, F., Zhang, S.: Research on wavelet neural network for fault location in power distribution network. Lect. Notes Electr. Eng. **138**, 1049–1057 (2012)
12. Dan-dan, M., Xiao-ru, W.: Single terminal methods of traveling wave fault location based on wavelet modulus maxima. Power Syst. Protect. Control **37**, 55–59 (2009)
13. Lu, H., Li, Y., Chen, M., Kim, H., Serikawa, S.: Brain intelligence: go beyond artificial intelligence. Mob. Netw. Appl. 1–10 (2017)
14. Lu, H., Li, Y., Mu, S., Wang, D., Kim, H., Serikawa, S.: Motor anomaly detection for unmanned aerial vehicles using reinforcement learning. IEEE Internet of Things J. https://doi.org/10.1109/JIOT.2017.2737479
15. Serikawa, S., Lu, H.: Underwater image dehazing using joint trilateral filter. Comput. Electr. Eng. **40**(1), 41–50 (2014)

Printed in the United States
By Bookmasters